THE ALGEBRAIC THEORY OF SPINORS

THE ALGEBRAIC THEORY OF SPINORS

By CLAUDE C. CHEVALLEY
PROFESSOR OF MATHEMATICS, COLUMBIA UNIVERSITY

MORNINGSIDE HEIGHTS, NEW YORK 1954
COLUMBIA UNIVERSITY PRESS

COLUMBIA BICENTENNIAL EDITIONS AND STUDIES

The Energetics of Development
BY LESTER G. BARTH AND LUCENA J. BARTH

New Letters of Berlioz, 1830–1868
TEXT WITH TRANSLATION, EDITED BY JACQUES BARZUN

On the Determination of Molecular Weights by Sedimentation and Diffusion
BY CHARLES O. BECKMANN AND OTHERS

LUIGI PIRANDELLO: *Right You Are*
TRANSLATED AND EDITED BY ERIC BENTLEY

The Sculpture of the Hellenistic Age
BY MARGARETE BIEBER

The Algebraic Theory of Spinors
BY CLAUDE C. CHEVALLEY

HENRY CARTER ADAMS: *Relation of the State to Industrial Action* AND *Economics and Jurisprudence*
EDITED BY JOSEPH DORFMAN

ERNST CASSIRER: *The Question of Jean-Jacques Rousseau*
TRANSLATED AND EDITED BY PETER GAY

The Language of Taxonomy
BY JOHN R. GREGG

Ancilla to Classical Reading
BY MOSES HADAS

JAMES JOYCE: *Chamber Music*
EDITED BY WILLIAM Y. TINDALL

Apocrimata: Decisions of Septimius Severus on Legal Matters
EDITED BY WILLIAM L. WESTERMANN AND A. ARTHUR SCHILLER

Published in Great Britain, Canada, India, and Pakistan by Geoffrey Cumberlege, Oxford University Press London, Toronto, Bombay, and Karachi

MANUFACTURED IN THE UNITED STATES OF AMERICA

Library of Congress Catalog Number: 54-5190

GENERAL EDITOR'S PREFACE

THE MODERN UNIVERSITY has become a great engine of public service. Its faculty of Science is expected to work for our health, comfort, and defense. Its faculty of Arts is supposed to delight us with plays and exhibits and to provide us with critical opinions, if not to lead in community singing. And its faculty of Political Science is called on to advise government and laity on the pressing problems of the hour. It is unquestionably right that the twentieth-century university should play this practical role.

But this conspicuous discharge of social duties has the effect of obscuring from the public—and sometimes from itself—the university's primary task, the fundamental work upon which all the other services depend. That primary task, that fundamental work, is Scholarship. In the laboratory this is called pure science; in the study and the classroom, it is research and teaching. For teaching no less than research demands original thought, and addressing students is equally a form of publication. Whatever the form or the medium, the university's power to serve the public presupposes the continuity of scholarship; and this in turn implies its encouragement. By its policy, a university may favor or hinder the birth of new truth. This is the whole meaning of the age-old struggle for academic freedom, not to mention the age-old myth of academic retreat from the noisy world.

Since these conditions of freedom constitute the main theme of Columbia University's Bicentennial celebration, and since the university has long been engaged in enterprises of public moment, it was doubly fitting that recognition be given to the activity that enlarges the world's "access to knowledge." Accordingly, the Trustees of the University and the Directors of its Press decided to signalize the 200th year of Columbia's existence by publishing some samples of its current scholarship. A full representation was impossible: limitations of time and space exercised an arbitrary choice. Yet the Bicentennial Editions and Studies, of which the titles are listed on a neighboring page, disclose the variety of products that come into being on the campus of a large university within a chosen year. From papyrology to the determination of molecular weights, and from the state's industrial relations to the study of an artist's or poet's work in its progress toward perfection, scholarship exemplifies the meaning of free activity, and seeks no other justification than the value of its fruits.

JACQUES BARZUN

CONTENTS

THE ALGEBRAIC THEORY OF SPINORS

INTRODUCTION

When E. Cartan classified the simple representations of all simple Lie algebras, he discovered a hitherto unknown representation of the orthogonal Lie algebra $\mathfrak{g}$, which could not be obtained from the representation on the vectors on which $\mathfrak{g}$ operates by the classical operations of constructing tensor products and decomposing them into simple (or irreducible) representation spaces. Cartan did not give a specific name to this representation; it was only later that, generalizing the terminology adopted in a special case by the physicists, he called the elements on which this new representation operates spinors. The simplest case of a spin representation is the one which presents itself for the orthogonal Lie algebra in 3 variables; this Lie algebra is well known to be isomorphic to the special unitary Lie algebra on 2 variables, which shows that it has a faithful representation of degree 2: this is its spin representation. Similarly, the fact that the orthogonal Lie algebra in 6 variables is isomorphic to the special unitary algebra in 4 variables reflects a special property of the spin representation of the first one of these algebras.

In his book, *Leçons sur la théorie des spineurs*,[1] Cartan recognized the connection between the spinors for a quadratic form Q and the maximal linear varieties of the quadratic cone of equation $Q = 0$. This connection is similar to the one which exists between subspaces of a vector space V and certain elements (the decomposable ones) of the exterior algebra over V: while every maximal linear variety on the cone $Q = 0$ is represented by a spinor, determined up to a scalar factor, not every spinor is correlated in this manner to a linear variety. Those which are we call "pure spinors"; in his book, Cartan indicates that it is possible to construct quadratic equations in the coefficients of an arbitrary spinor which give necessary and sufficient conditions for the spinor to be pure.

[1]E. Cartan, *Leçons sur la théorie de spineurs* (Paris: Hermann et Cie., 1938), 2 volumes.

The construction of the notion of spinor given by Cartan was rather complicated. In their paper,[2] R. Brauer and H. Weyl gave a much simpler presentation of the theory, based on the use of Clifford algebras. We follow their method in the present book, but we complete it by a simple construction of the pure spinors and of their relation with linear varieties on the cone $Q = 0$. In particular, we obtain a parametric representation of the pure spinors which is valid for all basic fields, while their characterization by the quadratic equations of Cartan breaks down for fields of characteristic 2.

The present book is oriented towards the algebraic and geometric applications of the theory of spinors; the author's lack of competence is the main reason for the complete absence of any application to physical theory. One of the most elegant purely mathematical applications is the one to the principle of triality in 8-dimensional space; we have devoted to it the last chapter of the present book, including a construction of the Cayley-Dickson algebra of octonions. We have not, however, included the description of the close connection which exists between the principle of triality on the one hand and, on the other hand, the exceptional Jordan algebra of dimension 27 and the five exceptional Lie groups; interesting as they are, these topics would have taken us too far away from the main subject of this book. In Chapter I, we establish those basic results in the theory of orthogonal groups which are to be of use in the remainder of the book; however, we have not included there the main result of the theory, namely, that the factor group of the commutator group of the orthogonal group by its center is simple when the index of the form is > 0; for this result we refer the reader to the book *Sur les groupes classiques* by J. Dieudonné.[3]

[2] R. Brauer and H. Weyl, "Spinors in *n* Dimensions," *American Journal of Mathematics*, *57* (1935), 425.

[3] J. Dieudonné, *Sur les groupes classiques* (Paris: Hermann et Cie., 1948).

PRELIMINARIES

1. Terminology

Throughout this book, with the exception of Section 5.2, we shall use the following conventions. The word "algebra" will mean "algebra with a unit element"; the symbol 1 will be used freely to denote the unit elements of the various algebras encountered (although unit elements may sometimes be denoted by specific symbols other than 1). We shall say that an algebra A is generated by a subset S of A when no proper subalgebra of A contains S and 1, i.e., when $\{1\} \cup S$ is a set of generators of A in the usual sense. By a homomorphism of an algebra A into an algebra B, we shall mean a homomorphism in the usual sense which, furthermore, maps the unit element of A upon that of B.

A representation of an algebra A (respectively: of a group G) on a vector space M is a homomorphism of A (respectively: G) into the algebra (respectively: group) of endomorphisms (respectively: automorphisms) of M. We say that ρ is *simple* if $M \neq \{0\}$ and if the only subspaces of M which are mapped into themselves by all operations of $\rho(A)$ (respectively: $\rho(G)$) are $\{0\}$ and M. If, in addition, it is true that the only endomorphisms of M which commute with all operations of $\rho(A)$ (respectively: $\rho(G)$) are the scalar multiples of the identity, then ρ is called *absolutely simple*. If M can be represented as a direct sum of subspaces $\neq \{0\}$, each of which is mapped into itself by the operations of $\rho(A)$ (respectively: $\rho(G)$), and is minimal with respect to this property, then ρ is called semi-simple. If this is the case, and M is finite-dimensional, then M may also be represented as the direct sum of subspaces $M_1, \cdots, M_h$ such that, for each i, the restrictions to M_i of the operations of $\rho(A)$ (respectively: $\rho(G)$) give a simple representation ρ_i of A (respectively: G). We shall then say that ρ is equivalent to the "sum" of the simple representations ρ_i, and we write $\rho \cong \rho_1 + \cdots + \rho_h$. If $h > 1$, then we say that ρ "splits" into the representations $\rho_1, \cdots, \rho_h$. If ρ' is any simple representation of A (respectively: G), then the number of indices i such that ρ_i is equivalent to ρ' is uniquely determined, and the sum of the spaces M_i relative to these indices is uniquely determined. In particular, if the representations ρ_i are all inequivalent to each other, then the spaces M_i are uniquely determined.

When s is an operation on a set M, we shall frequently denote by $s \cdot x$ (instead of $s(x)$) the transform by s of an $x \in M$.

2. Associative Algebras

We shall make a frequent use of the theory of finite-dimensional associative algebras; for an exposition of these results and their proofs, we refer the reader to a book by Jacobson.[1]

3. Exterior Algebras

We shall make use of a certain number of results on exterior algebras, which we indicate here.

Let M be a vector space over a field K, and E the exterior algebra[2] of M. For any $h \geq 0$, the products of h elements of M span a subspace E_h of E, and E is the direct sum of the spaces E_h ($h = 0, 1, \cdots$). The elements of E_h are called *homogeneous of degree* h; those among them which are representable as products of h elements of M are called *decomposable*. Assume now that M is of finite dimension m; then E_m is of dimension 1 and $E_h = \{0\}$ for $h > m$; if $(x_1, \cdots, x_m)$ is a base of M, then $x_1 \wedge \cdots \wedge x_m$ is a base of E_m. If $h \leq m$, then the products

$$x_{i_1} \wedge \cdots \wedge x_{i_h}, \qquad i_1 < \cdots < i_h \leq m,$$

form a base of E_h. *Any ideal $I \neq \{0\}$ of E contains E_m.* For, let $u = u_h + u_{h+1} + \cdots + u_m$ be an element $\neq 0$ of I, with $u_h \in E_h$, $u_h \neq 0$; write

$$u_h = \sum c(i_1, \cdots, i_h)x_{i_1} \wedge \cdots \wedge x_{i_h},$$

where $(i_1, \cdots, i_h)$ runs over the strictly increasing sequences of h integers between 1 and m, and let $(j_1, \cdots, j_h)$ be a sequence such that $c(j_1, \cdots, j_h) \neq 0$. Let $k_1, \cdots, k_{m-h}$ be all integers between 1 and m not occurring among $j_1, \cdots, j_h$; then it is easily seen that

$$x_{k_1} \wedge \cdots \wedge x_{k_{m-h}} \wedge u$$
$$= c(j_1, \cdots, j_h)x_{k_1} \wedge \cdots \wedge x_{k_{m-h}} \wedge x_{j_1} \wedge \cdots \wedge x_{j_h}$$

[1] N. Jacobson, *The Theory of Rings* (New York: The American Mathematical Society, 1943).

[2] See N. Bourbaki, *Eléments de mathématique*, Paris: Hermann et Cie., *Algèbre* Chapter III (1947); or C. Chevalley, *Théorie des groupes de Lie* (Paris: Hermann et Cie., 1951), II, Chapter I.

and the right side is an element $\neq 0$ of E_m because

$$(x_{k_1}, \cdots, x_{k_{m-h}}, x_{j_1}, \cdots, x_{j_h})$$

is a base of M.

Any linear mapping f of M into a vector space M' over K may be extended, in a unique manner, to a homomorphism F of E into the exterior algebra E' of M'; and F maps E_h into the space E'_h of homogeneous elements of degree h of E'. In particular, any endomorphism f of M may be extended to a homomorphism F of E into itself. If $\dim M = m$, $e \in E_m$, then we have $F(e) = (\det f)e$.

Let g be a linear form on M. Then there exists a uniquely determined antiderivation δ_g of E such that $\delta_g \cdot x = g(x)\cdot 1$ for all $x \in M$. The operation δ_g is homogeneous of degree -1; i.e., it maps any E_h into E_{h-1}, and 1 upon $\{0\}$. We have $\delta_g^2 = 0$; if g, g' are linear forms and a, a' scalars, then we have $\delta_{ag} = a\delta_g$, $\delta_{g+g'} = \delta_g + \delta_{g'}$, $\delta_g\delta_{g'} + \delta_{g'}\delta_g = 0$.

Let M^* be the dual space of M, and E^* the exterior algebra of M^*. Then, for each h, there exists a canonical bilinear form $(u, u^*) \rightarrow \langle u, u^*\rangle$ on $E_h \times E_h^*$, which defines an isomorphism of E_h^* with the, dual of E_h. Let s be any endomorphism of M, and ${}^ts$ the transpose of s, which is an endomorphism of M^* and maps any linear form $x^* \in M^*$ upon the linear form $x \rightarrow \langle x^*, s\cdot x\rangle = x^*(s\cdot x)$. Let S_h, S_h^* be the restriction to E_h, E_h^* of the homomorphisms of the algebras E, E^* into themselves which extend s, s^*; then we have $\langle u^*, S_h\cdot u\rangle = \langle S_h^*\cdot u^*, u\rangle$ for any $u \in E_h$, $u^* \in E^*_h$.

CHAPTER I

QUADRATIC FORMS

1.1. Bilinear Forms

Let M and N be vector spaces over the same field K. A bilinear form on $M \times N$ is by definition a mapping B of $M \times N$ into K with the following property: for every $x \in M$, the linear function $\lambda_x: y \to B(x, y)$ is a linear function on N; for every $y \in N$, the linear function $\mu_y: x \to B(x, y)$ is a linear function on M.

This being the case, we see immediately that the mapping $\lambda: x \to \lambda_x$ is a linear mapping of M into the dual space N^* of N, while the mapping $\mu: y \to \mu_y$ is a linear mapping of N into the dual space M^* of M. We shall say that λ and μ are the *linear mappings associated to B on the left and on the right.* Every linear mapping λ of M into N^* is associated to the left to a uniquely determined bilinear form B, given by

$$B(x, y) = (\lambda(x))(y).$$

Similarly, any linear mapping of N into M^* is associated to the right to a uniquely determined bilinear form.

Let P be a subspace of M. Then the set of elements $y \in N$ such that $B(x, y) = 0$ for all $x \in P$ is obviously a subspace P' of N, which is called the *right conjugate space* of P (with respect to B). Similarly, if Q is any subspace of N, the set of elements $x \in M$ such that $B(x, y) = 0$ for all $y \in Q$ is a subspace Q' of M, called the *left conjugate space* of Q. The following relations are obvious:

$$P \subset (P')' \text{ for any subspace } P \text{ of } M,$$

$$Q \subset (Q')' \text{ for any subspace } Q \text{ of } N,$$

$$(P_1 + P_2)' = P'_1 \cap P'_2 \text{ if } P_1, P_2 \text{ are subspaces of } M,$$

$$(Q_1 + Q_2)' = Q'_1 \cap Q'_2 \text{ if } Q_1, Q_2 \text{ are subspaces of } N.$$

The form B is called *nondegenerate* if we have $M' = \{0\}$, $N' = \{0\}$; this amounts to saying that the linear mappings λ, μ introduced above are one-to-one.

If B is any bilinear form on $M \times N$, and $x \in M$, $y \in N$, then the value of $B(x, y)$ depends only on the classes $\bar{x}$ of x modulo N' and $\bar{y}$ of y modulo M'; if we set $\bar{B}(\bar{x}, \bar{y}) = B(x, y)$, then $\bar{B}$ is obviously a nondegenerate bilinear form on the product $(M/N') \times (N/M')$.

Now assume that M and N are both finite-dimensional and denote their dimensions by m, n. If B is nondegenerate, then λ is an isomorphism of M with a subspace of N^*, and μ an isomorphism of N with a subspace of M^*. But N^* is of dimension n, and M^* of dimension m; it follows that $m \leq n$, $n \leq m$, whence $n = m$. Now, if we drop the assumption that B is nondegenerate, then we see that M/N' and N/M' have the same dimension; their dimension is called the *rank* of the bilinear form B.

I.1.1. *Let B be a nondegenerate bilinear form on the product of two m-dimensional vector spaces M and N. If P is a p-dimensional subspace of M, then its conjugate P' is of dimension $m - p$, and $(P')' = P$; if Q is a q-dimensional subspace of N, then Q' is of dimension $m - q$ and $(Q')' = Q'$.*

The linear mapping λ associated to the left to B is an isomorphism of M with the dual N^* of N, and P' is the set of solutions of the linear equations $\lambda_x(y) = 0$ for all $x \in P$. Since λ maps P upon a p-dimensional subspace of N^*, P' is of dimension $m - p$. We prove in the same way that Q' is of dimension $m - q$; in particular, $(P')'$ is of dimension $m - (m - p) = p$ and contains P, whence $(P')' = P$; we see in the same way that $(Q')' = Q$.

We shall be mainly interested in bilinear forms B on the product $M \times M$ of a finite-dimensional vector space M by itself. Let $(x_1, \cdots, x_m)$ be a base of M, and set $b_{ij} = B(x_i, x_j)$. Then, clearly, we have

$$B\left(\sum_{i=1}^{m} a_i x_i, \sum_{i=1}^{m} a'_i x_i\right) = \sum_{i,j=1}^{m} b_{ij} a_i a'_j .$$

The matrix $B = (b_{ij})$ is called the *matrix of the form* B with respect to the base $(x_1, \cdots, x_m)$; its determinant is called the *discriminant* of B with respect to the base $(x_1, \cdots, x_m)$. It is clear that any $(n \times n)$-square matrix with coefficients in K is the matrix of some uniquely determined bilinear form on $M \times M$. Let $(x^*_1, \cdots, x^*_m)$ be the base of the dual M^* of M dual to the base $(x_1, \cdots, x_m)$ of M. The notation being as above, we have

$$\mu_{x_j}(x_i) = b_{ij} ,$$

whence

$$\mu_{x_j} = \sum_{i=1}^{m} b_{ij}x_i^*;$$

B is therefore also the matrix which represents μ with respect to the bases $(x_1, \cdots, x_m)$ of M and $(x_1^*, \cdots, x_m^*)$ of M^*. Since the rank r of B is the rank of the linear mapping μ, r is equal to the rank of **B**. Then, a necessary and sufficient condition for B to be nondegenerate is for its discriminant (with respect to any base) to be $\neq 0$.

Let $(y_1, \cdots, y_m)$ be any other base of M. Write

$$y_i = \sum_{j=1}^{m} t_{ji}x_j ,$$

and let **T** be the matrix (t_{ij}). Then we see immediately that the matrix of B with respect to the base $(y_1, \cdots, y_m)$ is ${}^t\mathbf{T}.\mathbf{B}.\mathbf{T}.$, where ${}^t\mathbf{T}$ is the transpose of **T**. Its determinant is the product $(\det \mathbf{B})(\det \mathbf{T})^2$.

We shall be interested mainly in bilinear forms B which are *symmetric*, i.e., such that

$$B(x, y) = B(y, x)$$

for any x, y in M. Moreover, in the case where K is of characteristic 2, we shall be interested only in those bilinear forms B for which

$$B(x, x) = 0$$

for all $x \in M$. Such a form is usually called *alternating*. In the case of characteristic 2, the condition of being alternating implies the symmetry, for the relations $B(x, x) = B(y, y) = B(x + y, x + y) = 0$ imply $B(x, y) + B(y, x) = 0$. *We shall make the convention in that case to call symmetric only those bilinear forms which are alternating.*

If B is symmetric, the distinction between left and right conjugates disappears, and we shall therefore simply speak of the *conjugate* of a subspace of M.

A subspace P of M is called *isotropic* if it has an element $\neq 0$ in common with its conjugate P', and *totally isotropic* if $P \subset P'$. To say that P is isotropic is to say that the restriction B_P of B to $P \times P$ is degenerate; to say that P is totally isotropic is to say that $B_P = 0$. An element $x \in M$ is called isotropic if $B(x, x) = 0$. If P is a subspace of M, every element of $P \cap P'$ is isotropic; if K is of characteristic 2, every element of M is isotropic.

I.1.2. *Let B be a nondegenerate symmetric bilinear form on $M \times M$, M of finite dimension. If P is a nonisotropic subspace of M, P' is nonisotropic and M is the direct sum of P and P'.*

Let m and p be the dimensions of M and P; then P' is of dimension $m - p$. We have $P \cap P' = \{0\}$ and $(P')' = P$, which shows that P' is not isotropic; since $\dim P + \dim P' = \dim M$ and $P \cap P' = \{0\}$, M is the direct sum of P and P'.

1.2. Quadratic Forms

Let M be a vector space over a field K. A quadratic form on M is by definition a mapping Q of M into K which has the following properties:

(a) $$Q(ax) = a^2Q(x) \qquad (a \in K, x \in M).$$

(b) The mapping $(x, y) \rightarrow Q(x + y) - Q(x) - Q(y)$ is a bilinear form B on $M \times M$.

We shall say that B is the *bilinear form associated to* Q. It is clear from the definition that $B(y, x) = B(x, y)$ for any $x, y \in M$. Moreover, we have $Q(2x) = 4Q(x)$ by (a), whence, by (b),

$$B(x, x) = 2Q(x).$$

It follows that $B(x, x) = 0$ if K is of characteristic 2; B is therefore symmetric.

If there is given a quadratic form Q on M, we shall call conjugate of a subspace P of M the conjugate of P relative to the bilinear form B associated to Q; and we define similarly the notions of isotropic spaces, totally isotropic spaces, and isotropic elements.

The restriction of the mapping Q to a subspace N of M is a quadratic form on N whose associated bilinear form is the restriction of B to $N \times N$. If this restriction is the zero quadratic form on N, then we say that N is *totally singular*. Any $x \in M$ such that $Q(x) = 0$ is a called *singular*.

I.2.1. *Any totally singular subspace N of M is totally isotropic. If the characteristic of K is $\neq 2$, any totally isotropic subspace is totally singular.*

If $x, y \in N$, we have $x + y \in N$ and

$$B(x, y) = Q(x + y) - Q(x) - Q(y) = 0,$$

which proves the first assertion. The second one follows immediately from the formula $B(x, x) = 2Q(x)$.

Let M' be the conjugate of the whole space M. If K is not of characteristic 2, M' is totally singular. If K is of characteristic 2, we have $Q(x + y) = Q(x) + Q(y)$ for $x, y \in M'$, from which it follows immediately that the set M'_0 of singular vectors contained in M' is a vector subspace of

M'. Assume that M is finite-dimensional; let m, m', m'_0 be the dimensions of M, M', M'_0. Then $m - m'$ is the rank of B. The number $m - m'_0$ is called the *rank* of Q, and $m' - m'_0$ is called its *defect*. The defect is 0 if K is not of characteristic 2.

We shall now assume that M is finite dimensional.

I.2.2. *Let B_0 be any bilinear form on $M \times M$. Then $x \to B_0(x, x)$ is a quadratic form on Q, and any quadratic form may be represented in this manner.*

We have $B_0(ax, ax) = a^2B_0(x, x)$ if $a \in K$, and $B_0(x + y, x + y) - B_0(x, x) - B_0(y, y) = B_0(x, y) + B_0(y, x)$, which proves the first assertion. Now, let Q be any quadratic form on M, B its associated bilinear form, and $(x_1, \cdots, x_m)$ a base of M. Let $b_{ij} = B(x_i, x_j)$; then, clearly, we have

$$Q\left(\sum_{i=1}^{m} a_i x_i\right) = \sum_{i=1}^{m} a_i^2 Q(x_i) + \sum_{i<j} a_i a_j b_{ij} . \tag{1}$$

We may define a bilinear form B_0 on $M \times M$ by the formula

$$B_0\left(\sum_{i=1}^{m} a_i x_i , \sum_{i=1}^{m} a'_i x_i\right) = \sum_{i=1}^{m} a_i a'_i Q(x_i) + \sum_{i<j} a_i a'_j b_{ij} .$$

It is then clear that $Q(x) = B_0(x, x)$.

If K is not of characteristic 2, we may take, in I.2.2, $B_0 = \frac{1}{2} B$.

I.2.3. *Let K' be an overfield of K, $M^{K'}$ the vector space over K' deduced from M by extending the basic field to K' and Q a quadratic form on M. Then there exists a uniquely determined quadratic form on $M^{K'}$ which extends Q. Its associated bilinear form is an extension of that of Q.*

Let B_0 be a bilinear form on $M \times M$ such that $Q(x) = B_0(x, x)$. Then B_0 may be extended to a bilinear form B'_0 on $M^{K'} \times M^{K'}$. For, let $(x_1, \cdots, x_m)$ be a base of M; then it suffices to take for B'_0 the bilinear form on $M^{K'} \times M^{K'}$ whose matrix with respect to $(x_1, \cdots, x_m)$ is the same as that of B_0. The formula $Q'(x) = B'_0(x, x)$ then defines a quadratic form on $M^{K'}$ which extends Q. If a quadratic form Q'_1 on $M^{K'}$ extends Q, its associated bilinear form clearly extends B. Making use of formula (1) above, applied to Q'_1, where $a_1, \cdots, a_m$ are allowed to run over K', we see that there exists only one quadratic form over $M^{K'}$ which extends Q.

Two quadratic forms Q, Q_1 on vector spaces M, M_1 over K are called *equivalent* when there is an isomorphism σ of M with M_1 such that

$Q_1(\sigma \cdot x) = Q(x)$ for all $x \varepsilon M$. This clearly implies that $B_1(\sigma \cdot x, \sigma \cdot y) = B(x, y)$ for any $x, y \varepsilon M$, if B, B_1, are the associated bilinear forms to Q, Q_1.

1.3. Special Bases

We shall denote by Q a quadratic form on a finite dimensional vector space M, and by B the bilinear form associated to Q. Two vectors x, y in M are called *orthogonal* to each other if $B(x, y) = 0$.

I.3.1. *If K is not of characteristic 2, then M has a base composed of mutually orthogonal vectors.*

We prove this by induction on the dimension m of M. If $m = 0$, there is nothing to prove. Assume that the statement is true for spaces of dimension $m - 1$. If $Q = 0$, then the statement is trivial. Assume that $Q \neq 0$, and let x_1 be a vector such that $Q(x_1) \neq 0$; let N be the conjugate space of Kx_1. It is clear that N is of dimension $\geq m - 1$; since $Q(x_1) \neq 0$ and K is not of characteristic 2, $B(x_1, x_1) \neq 0$, and x_1 is not in N. We conclude that M is the direct sum of Kx_1 and N, and that dim $N = m - 1$. Thus, there is a base $(x_2, \cdots, x_m)$ of N composed of mutually orthogonal vectors. Then $(x_1, x_2, \cdots, x_m)$ is a base of M composed of mutually orthogonal vectors.

Any base of M whose vectors are mutually orthogonal is called an *orthogonal base.*

I.3.2. *Assume that the bilinear form B is nondegenerate. Let N be a totally isotropic subspace of M of dimension r. Then there exists a totally isotropic subspace P of dimension r such that $N \cap P = \{0\}$ and $N + P$ is not isotropic. If $(x_1, \cdots x_r)$ is a base of N, and P has the properties stated above, there is a base $(y_1, \cdots, y_r)$ of P such that $B(x_i, y_j) = \delta_{ij}$, $(1 \leq i, j \leq r)$. If N is totally singular, then P may be taken to be totally singular. Let R be the conjugate space of $N + P$. If N is maximal in the set of totally singular subspaces, then we have $Q(x) \neq 0$ for every $x \neq 0$ in R.*

Let p be any integer ≥ 0 and $< r$; suppose that we have already constructed p vectors $y_1, \cdots, y_p$ with the following properties: $B(x_i, y_j) = \delta_{ij}$ for $1 \leq i \leq r$, $1 \leq j \leq p$; the space spanned by $y_1, \cdots, y_p$ is totally isotropic and is totally singular in case N is. The conjugate of the space spanned by the x_i's for $i \neq p + 1$ is of dimension $m - r + 1$, if $m = \dim M$ (by I.1.1.), while the conjugate of N is of dimension $m - r$. Thus, there is a vector y such that $B(y, x_i) = 0$ for $i \neq p + 1$, $B(y, x_{p+1}) \neq 0$, and we may obviously assume that $B(y, x_{p+1}) = 1$.

Since N is totally isotropic, any $y' \in y + N$ has the same properties as y. Let $b_i = B(y, y_i)$ for $i \leq p$ and

$$y' = y - \sum_{i=1}^{p} b_i x_i \;;$$

then y' has the same properties as y and is further orthogonal to $y_1, \cdots, y_p$. If $c \in K$, then we have

$$B(y' - cx_{p+1}, y' - cx_{p+1}) = B(y', y') - 2c,$$

$$Q(y' - cx_{p+1}) = Q(y') + c^2Q(x_{p+1}) - c.$$

If K is of characteristic $\neq 2$, then we may choose c such that $y' - cx_{p+1}$ is isotropic. If K is of characteristic 2, any vector is isotropic. Thus, we may always select c in such a way that $y_{p+1} = y' - cx_{p+1}$ is isotropic. If K is not of characteristic 2, this implies that $Q(y_{p+1}) = 0$. If K is of characteristic 2 and N totally singular, then $Q(x_{p+1}) = 0$ and we may take c such that $Q(y_{p+1}) = 0$. It is clear that

$$B(x_i, y_{p+1}) = \delta_{i,p+1} \qquad (1 \leq i \leq r)$$

and that

$$B(y_i, y_j) = 0 \qquad (1 \leq i, j \leq p + 1),$$

which shows that the space spanned by $y_1, \cdots, y_{p+1}$ is totally isotropic. If N is totally singular, then $Q(y_i) = 0$ $(1 \leq i \leq p + 1)$, and the space spanned by $y_1, \cdots, y_{p+1}$ is totally singular.

At the end of this construction, we obtain r vectors $y_1, \cdots, y_r$ such that $B(x_i, y_j) = \delta_{ij}$ $(1 \leq i, j \leq r)$ and the space $P = Ky_1 + \cdots + Ky_r$ is totally isotropic; moreover, P is totally singular if N is. We have

$$B_i\left(x, \sum_{j=1}^{r} a_j y_j\right) = a_j \qquad (1 \leq j \leq r),$$

which implies that $y_1, \cdots, y_r$ are linearly independent and that P has only 0 in common with the conjugate N' of N; this in turn obviously implies that $N + P$ is not isotropic. Let R be its conjugate; then M is the direct sum of $N + P$ and R (I.1.2). If N is totally singular and R contains a $z \neq 0$ such that $Q(z) = 0$, then we have $B(z, x) = 0$ for every $x \in N$, whence $Q(x + z) = Q(x) + Q(z) + B(z, x) = 0$ and $N + Kz$ is totally singular. This concludes the proof of I.3.2.

I.3.3. *Assume that B is nondegenerate and that there is an $x \neq 0$ in M such that $Q(x) = 0$. Then for any $a \in K$, there is a $z \in M$ such that $Q(z) = a$.*

It follows from I.3.2, applied to $N = Kx$, that there is a $y \in M$ such that $Q(y) = 0$, $B(x, y) = 1$. We then have $Q(x + ay) = a$.

I.3.4. *Assume K is algebraically closed. Denote by m the dimension of M and by N a maximal totally singular subspace of M. If B is nondegenerate, N is of dimension $[m/2]$ (integral part of $m/2$). If we assume further that K is of characteristic 2, then m is even.*

The notation being as I.3.2, assume that N is maximal totally singular. Let z, z' be vectors in R; since K is algebraically closed, we can find elements a, a' not both 0 of K such that

$$Q(az + a'z') = a^2Q(z) + aa'B(z, z') + a'^2Q(z') = 0.$$

It follows that $az + a'z' = 0$, and that R is of dimension 0 or 1, whence $m = 2r$ or $m = 2r + 1$. Assume that $m = 2r + 1$, and let z be an element $\neq 0$ of R. Since z belongs to the conjugate of $N + P$ but not to that of M, we have $B(z, z) \neq 0$, and K is not of characteristic 2.

Still assuming that K is algebraically closed, we see that, if m is even, M has a base $(x_1, \cdots, x_r, y_1, \cdots, y_r)$ such that

$$Q\left(\sum_{i=1}^{r} (a_i x_i + b_i y_i)\right) = \sum_{i=1}^{r} a_i b_i\,, \tag{1}$$

while, if m is odd, M has a base $(x_1, \cdots, x_r, y_1, \cdots, y_r, z)$ such that

$$Q\left(\sum_{i=1}^{r} (a_i x_i + b_i y_i) + cz\right) = \sum_{i=1}^{r} a_i b_i + c^2. \tag{2}$$

These results are valid under the assumption that Q is of maximal rank m equal to the dimension of M and has defect 0 if K is of characteristic 2.

I.4. The Orthogonal Group

We shall denote by Q a quadratic form on a finite-dimensional vector space M over a field K; we shall assume that the associated bilinear form B of Q is nondegenerate.

A linear mapping s of M into itself is called *orthogonal* (relative to Q) if we have $Q(s\cdot x) = Q(x)$ for all $x \in M$. It follows immediately that $B(s\cdot x, s\cdot y) = B(x, y)$ for all $x, y \in M$. Thus, if $s\cdot x = 0$, then we have $B(x, y) = 0$ for every $y \in M$, whence $x = 0$; this shows that any orthogonal mapping is an automorphism of M. It is clear that the orthogonal mappings form a group; this group is called the *orthogonal group* of Q and will be denoted by G.

A vector-space isomorphism s of a subspace N of M with a subspace P is called a Q-isomorphism if $Q(s\cdot x) = Q(x)$ for every $x \in N$; this implies that $B(s\cdot x, s\cdot y) = B(x, y)$ for $x, y \in N$.

I.4.1. *The assumptions and notation being as stated above, every Q-isomorphism of a subspace N of M with a subspace P may be extended to an operation of the group G.*

We proceed by induction on the dimension n of N. Our statement is obvious for $n = 0$. Assume that $n > 0$ and that the statement is true for subspaces of dimension $n - 1$. Let U be an $(n - 1)$-dimensional subspace of N. The restriction of s to U may be extended to an operation $s_0 \in G$. Let $s'(x) = s_0^{-1}(x)s(x)$ for $x \in N$. Then s' is a Q-isomorphism of N which leaves the elements of U fixed. If s' extends to an operation $s'_0 \in G$, then $s_0 s'_0$ is an element of G which extends s. Thus, we see that we may assume without loss of generality that s leaves the elements of U fixed. Let $\mathfrak{B}$ be the set of subspaces V of M with the following property: s may be extended to a Q-isomorphism of $V + N$, leaving the elements of V fixed. Let V_1 be a maximal element of $\mathfrak{B}$, and s_1 the Q-isomorphism of $N_1 = V_1 + N$ which extends s and leaves the elements of V_1 fixed. Let $P_1 = s_1(N_1)$, $U_1 = V_1 + U$; if $U_1 = N_1$, then s_1 is the identity and the statement is obvious. If not, let x_1 be an element of N_1 not in U_1, and $y_1 = s_1 \cdot x_1$, whence $N_1 = U_1 + Kx_1$, $P_1 = U_1 + Ky_1$, $Q(x_1) = Q(y_1)$.

Assume that we have elements $z, z' \in M$ with the following properties: z is not in N_1, z' is not in P_1, $z' - z$ is in the conjugate space U_1' of U_1, $B(z', y_1) = B(z, x_1)$, $Q(z) = Q(z')$. Then we may extend s_1 to an isomorphism s_2 of $N_1 + Kz$ with $P_1 + Kz'$ which maps z upon z'. We shall see that s_2 is a Q-isomorphism. Any $x \in N_1$ is of the form $u + ax_1$, $u \in U_1$, $a \in K$, and $s_1 \cdot x = u + ay_1$. Since $B(z' - z, u) = 0$, $B(z, x_1) = B(z', y_1)$, we have $B(z, x) = B(z', s_1 \cdot x)$; on the other hand, we have $Q(x) = Q(s_1 \cdot x)$, whence, for $b \in K$,

$$\begin{aligned} Q(bz + x) &= b^2Q(z) + bB(z, x) + Q(x) \\ &= b^2Q(z') + bB(z', s_1 \cdot x) + Q(s_1 \cdot x) = Q(bz + s_1 \cdot x), \end{aligned}$$

which proves that s_2 is a Q-isomorphism.

Let H be the conjugate of the space $K(x_1 - y_1)$; if $z \in H$, then we have $B(z, x_1) = B(z, y_1)$. Applying the above considerations with $z' = z$, we see that it follows from the maximal character of V_1 that z lies in N_1 or in P_1, whence $H = (H \cap N_1) \cup (H \cap P_1)$. Were $H \cap N_1$ and $H \cap P_1$ both $\neq H$, then there would exist elements $z_1 \in H \cap N_1$, $z'_1 \in H \cap P_1$ such that z_1 is not in $H \cap P_1$ and z'_1 not in $H \cap N_1$; $z = z_1 + z'_1$ would then be an element of H not belonging to $N_1 \cup P_1$, which is impossible. Thus, H is contained in one of the spaces N_1, P_1. If $N_1 = M$, then we are through. If not, we see that H, which is of dimension dim $M - 1$, is identical with one of the spaces N_1 or P_1, which shows that $x_1 - y_1$ is orthogonal to at least one of x_1, y_1. But we have $B(x_1, x_1) = B(y_1, y_1)$; thus, $B(x_1, x_1 - y_1) = B(y_1 - x_1, y_1)$ and both x_1 and y_1 are in H. It follows immediately that $N_1 = P_1 = H$. Let z be an element of M not in H, whence $B(z, x_1 - y_1) \neq 0$. Then it is clear that $M = H + Kz = N_1 + Kz$. We shall construct an element z' with the properties stated above; s_2 will then be an operation of G extending s. It is clear that y_1 is not in U_1; the conjugate of U'_1 being U_1, U'_1 contains a vector which is not orthogonal to y_1, and therefore also a vector u such that $B(u, y_1) = B(z, x_1 - y_1)$. Since $B(x_1 - y_1, y_1) = 0$, u is not in $K(x_1 - y_1)$; i.e., u is not in the conjugate of H; since $u \in U'_1$, $H = N_1 = U_1 + Kx_1$, we have $B(u, x_1) \neq 0$ and $B(z + u, x_1 - y_1) = B(u, x_1) \neq 0$, which shows that $z + u$ is not in $P_1 = H$. Let c be any element of K; since $x_1 - y_1 \in P_1$, $z + u + c(x_1 - y_1)$ is not in P_1. Since $x_1 - y_1$ is in the conjugate of H and $U_1 \subset H$, $(z + u + c(x_1 - y_1)) - z$ is in U'_1. We have

$$Q(x_1 - y_1) = Q(x_1) + Q(y_1) - B(x_1, y_1) = 2Q(x_1) - B(x_1, y_1)$$
$$= B(x_1, x_1) - B(x_1, y_1) = 0.$$

It follows that

$$Q(z + u + c(x_1 - y_1)) = Q(z + u) + cB(z + u, x_1 - y_1).$$

Since $B(z + u, x_1 - y_1) \neq 0$, c may be determined in such a way that $Q(z + u + c(x_1 - y_1)) = Q(z)$. If we set $z' = z + u + c(x_1 - y_1)$, z and z' have the required properties, and I.4.1 is proved.

I.4.2. *Let N be a totally singular subspace of M. Then every automorphism of N may be extended to an operation of G.*

This follows immediately from I.4.1.

I.4.3. *All maximal totally singular subspaces of M have the same dimension and are permuted transitively among themselves by the operations of G.*

It follows immediately from I.4.1 that, if N and P are totally singular subspaces of the same dimension, there is an operation s of G which transforms N into P. If P' is a totally singular space containing P, $s^{-1}(P')$ is a totally singular space containing N. Assume that N is maximal totally singular of dimension r; every subspace of a totally singular space being totally singular, it is clear that there cannot exist any totally singular subspace of dimension $> r$ in M.

The common dimension of all totally singular maximal subspaces of M is called the *index* of Q. Let r be its value. Then it follows from I.3.2 that $r \leq [m/2]$. Moreover, if $r = [m/2]$, there is a base of M with respect to which Q has one of the expressions (1), (2) of I.3; if K is of characteristic 2, m is necessarily even.

I.4.4. *The notation and assumptions being as above, let s be an operation of G, L the set of elements of M left fixed by s, and u the linear mapping $x \to s \cdot x - x$ of M into itself. Then $u(M)$ is the conjugate space of L.*

If $x \in M$, $y \in L$, then $B(y, s \cdot x) = B(s \cdot y, s \cdot x) = B(y, x)$ and y is orthogonal to $s \cdot x - x$. Let ν be the dimension of L; since L is the kernel of u, $u(M)$ is of dimension $m - \nu$ (where $m = \dim M$); being contained in the conjugate of L, it is identical to it.

I.4.5. *The notations and assumptions being as above, assume further that Q is of index $m/2 = r$, and let N be a maximal totally singular subspace of M. Let H be the group of orthogonal mappings which leave all points of N fixed. If $x \in M$, $s \in H$, then $s \cdot x - x$ belongs to N. Let P be a totally singular subspace of M such that $M = N + P$. If $s \in H$, $y \in P$, $y' \in P$, set $\Gamma_s(y, y') = B(y, s \cdot y')$; then Γ_s is an alternating bilinear form on $P \times P$, and $s \to \Gamma_s$ is an isomorphism of H with the additive group of all alternating bilinear forms on $P \times P$. The rank of Γ_s is the dimension of the image of M under the mapping $x \to s \cdot x - x$. If s, s' are elements of H such that Γ_s and $\Gamma_{s'}$ have the same rank, then s and s' are conjugate to each other in G.*

If $x \in M$, $s \cdot x - x$ is in the conjugate N' of N by I.4.4; but N' contains N and is of dimension $m - r = r$, whence $N' = N$ and $s \cdot x - x \in N$. The function Γ_s is obviously bilinear. If $y \in P$, then we have $0 = Q(y) = Q(s \cdot y) = Q(y + (s \cdot y - y)) = B(y, s \cdot y - y) = \Gamma_s(y, y)$, since $Q(s \cdot y - y) = 0$; it follows that Γ_s is alternating. If s, s' are in H, then $s \cdot (s' \cdot y - y) = s' \cdot y - y$ and therefore we have $ss' \cdot y - y = (s \cdot y - y) + (s' \cdot y - y)$, $\Gamma_{ss'} = \Gamma_s + \Gamma_{s'}$. If $\Gamma_s = 0$, then for any $y' \in P$, $s \cdot y' - y'$ is in N and also in the conjugate of P, which is P, whence $s \cdot y' = y'$, and s is the identity. Conversely, let Γ be any alternating bilinear form on $P \times P$. For any $x \in N$, let λ_x be the linear form $y \to B(x, y)$ on P; then $x \to \lambda_x$ is a linear mapping of N into the dual P^* of P. We have $B(x, x') = 0$ if $x' \in N$; thus, $\lambda_x = 0$ implies $B(x, z) = 0$ for all $z \in M$, whence $x = 0$, and $x \to \lambda_x$ is an isomorphism of N with P^*. It follows that, for every $y' \in P$, there is a unique $u(y') \in N$ such that $B(y, u(y')) = \Gamma(y, y')$ for all $y \in P$; u is obviously a linear mapping of P into N. The formula $s(x + y) = x + y + u(y)$ $(x \in N, y \in P)$ defines an automorphism of the vector space M. Since $Q(x) = Q(y) = Q(u(y)) = 0$, $B(x, u(y)) = 0$, $B(y, u(y)) = \Gamma(y, y) = 0$, we have $Q(x + y + u(y)) = B(x, y) = Q(x + y)$ and $s \in G$. It is clear that $\Gamma_s = \Gamma$. The rank of Γ is even; let it be 2ρ. Then it is well known that there is a base $(y_1, \cdots, y_r)$ of P such that $\Gamma(y_i, y_j) = 1$ if $i = 2k - 1, j = 2k, k \leq \rho$, -1 if $i = 2k, j = 2k - 1, k \leq \rho$ and 0 otherwise. Let $x_{2k-1} = u(y_{2k})$, $x_{2k} = -u(y_{2k-1})$ if $k \leq \rho$; then we have $B(x_i, y_j) = \delta_{ij}$ if $1 \leq i, j \leq 2\rho$. It follows easily that we may include $x_1, \cdots, x_{2\rho}$ in a base $(x_1, \cdots, x_r)$ of N such that $B(x_i, y_j) = \delta_{ij}$ $(1 \leq i, j \leq r)$. We have

$$s \cdot y_{2k-1} = y_{2k-1} - x_{2k}, \qquad s \cdot y_{2k} = y_{2k} + x_{2k-1} \qquad (k \leq \rho)$$

and $s \cdot y_i = y_i$ if $i > \rho$. Now, let s' be an operation of H such that $\Gamma_{s'}$ is of rank 2ρ; let $(x'_1, \cdots, x'_r, y'_1, \cdots, y'_r)$ be determined from $\Gamma_{s'}$ as $(x_1, \cdots, x_r, y_1, \cdots, y_r)$ have been from Γ. Since $Q(x_i) = Q(x'_i) = Q(y_i) = Q(y'_i) = 0$, $B(x_i, y_j) = B(x'_i, y'_j)$, the automorphism t of M which maps x_i upon x'_i and y_i upon y'_i $(1 \leq i \leq r)$ is in G, and it is clear that $tst^{-1} = s'$.

1.5. Symmetries

We denote by Q a quadratic form on a vector space M of finite dimension m over a field K; we assume that the associated bilinear form B of Q is nondegenerate. We denote by G the orthogonal group of Q.

Let H be a hyperplane whose conjugate contains a nonsingular vector z. Let $Q(z) = a$, and, for $x \in M$,

$$s \cdot x = x - a^{-1}B(x, z)z.$$

Then s is an endomorphism of M, and an easy computation shows that $Q(s \cdot x) = Q(x)$; i.e., s is orthogonal. It is clear that s does not change if we replace z by kz, $k \neq 0$; the conjugate of H being Kz, s depends only on H and is called the *symmetry with respect to* H. It is clear that s leaves the points of H and only these invariant; since $B(z, z) = 2Q(z)$, we have $s \cdot z = -z$. The operation s is the only orthogonal operation distinct from the identity which leaves the points of H fixed. For, let s' be an operation with these properties. Clearly, if x is not in H, $s' \cdot x \neq x$ and $s' \cdot x - x$ is in the conjugate of H (by I.4.4), whence $s' \cdot x = x + kz$. Since $Q(s' \cdot x) = Q(x)$, we have $kB(x, z) + k^2a = 0$, and, since $k \neq 0$, we have $k = -a^{-1} B(x, z)$, whence $s' = s$. If $t \in G$, then, clearly, tst^{-1} is the symmetry with respect to the hyperplane $t(H)$.

I.5.1 (Cartan, Dieudonné). *Except in the case where K has only 2 elements, M is of dimension 4 and Q of index 2, every operation of G*

belongs to the group G' generated by the symmetries with respect to the hyperplanes whose conjugates contain nonsingular vectors.

If $s \in G$, denote by $L(s)$ the set of fixed points of s, by $\nu(s)$ its dimension, and by u_s the linear mapping $x \to s \cdot x - x$. Assume that $u_s(M)$ contains a nonsingular vector $z = s \cdot y - y$, and let t be the symmetry with respect to the conjugate hyperplane H of Kz. Then we have

$$Q(z) = Q(s \cdot y) + Q(y) - B(s \cdot y, y) = 2Q(y) - B(s \cdot y, y)$$
$$= B(y, y) - B(s \cdot y, y) = -B(z, y)$$

and $t \cdot y = y + z = s \cdot y$, whence $t(s \cdot y) = y$, and $y \in L(ts)$. On the other hand, $L(s) \subset H$ (by I.4.4), whence $L(s) + Ky \subset L(ts)$ and $\nu(ts) > \nu(s)$. Now assume that s' is one of the elements of the coset $G's$ for which $\nu(s')$ is the largest possible: then we see that $u_{s'}(M)$ is totally singular. Let us call singular those $s \in G$ for which $u_s(M)$ is totally singular; if s is singular, we call index of s the dimension $\rho(s) = m - \nu(s)$ of $u_s(M)$.

Now we shall prove that any two maximal totally singular spaces N, N' may be transformed into each other by an operation of G'. It is clearly sufficient to prove that, if $N \cap N'$ is of dimension $l < \dim N$, there exists a hyperplane H whose conjugate contains a nonsingular vector such that the symmetry t with respect to H transforms N' into a space $t(N')$ such that $\dim (N \cap t(N')) > l$. Since $\dim (N + N') > \dim N$, $N + N'$ contains a nonsingular vector $z = x + x'$ ($x \in N$, $x' \in N'$). Since $Q(z) = B(x, x') \neq 0$, x does not belong to N'. We take for H the conjugate of Kz; then we have

$$t(x') = x' - (B(x, x'))^{-1}B(x, x')z = -x \in N.$$

On the other hand, if $x'' \in N \cap N'$, we have $B(x, x'') = B(x', x'') = B(z, x'') = 0$ and $x'' \in t(N')$; thus, we have $N \cap t(N') \supset N \cap N' + Kx$, which proves our assertion.

This being said, let s be a singular operation of G; then $u_s(M)$ is contained in a maximal totally singular space N; since the conjugate of N is in the conjugate of $u_s(M)$, it is in $L(s)$. Let H_N be the group of operations of G which leave fixed all points of the conjugate of N. There is a maximal totally singular space P such that $N + P$ is not isotropic (I.3.2); let R be the conjugate of this space. If $r = \dim N$, then the conjugate of N, which is of dimension $m - r$, contains $N + R$, also of dimension $m - r$; this conjugate is therefore $N + R$. Any Q-automorphism of $N + P$ may obviously be extended to an operation of G, leaving the points of R fixed. Thus, it follows from I.4.5 that H_N is an abelian group which is isomorphic under a mapping $s \to \Gamma_s$ to the additive group of alternating bilinear forms on $P \times P$, and the rank of Γ_s is the index of s. Thus, two operations of same index of H_N are conjugate in G.

Every singular operation s' may be transformed into an operation of H_N by some operation of G'. For, let N' be a maximal totally singular space containing $u_{s'}(M)$ and t an operation of G' such that $t(N') = N$. Since s' leaves the elements of the conjugate of N' fixed, $ts't^{-1}$ is in H_N.

Now, it is clear that G' is a normal subgroup of G. It follows from what we have just said that $G = H_N G'$; G/G' is therefore abelian. Moreover, if s, s' are singular operations with the same index, then they are conjugate in G, which shows that their classes $\bar{s}$, $\bar{s}'$ modulo G' are equal. Thus, if there are singular operations s, s' such that s, s', and ss' have the same index, then $\bar{s} = \bar{s}' = \bar{s}\bar{s}'$, and $s, s' \in G'$. If K has more than 2 elements, let a be $\neq 0, -1$ in K, and $s \in H_N$. Then Γ_s, $a\Gamma_s$, and $(1 + a)\,\Gamma_s$ have the same rank, whence $s \in G'$. If K has two elements, let r be the index of Q; if $r = 0$ or 1, then H_N contains only the identity (the rank of any alternating bilinear form being even). Assume that $r > 2$, $m > 4$. Every alternating form is obviously representable as a sum of forms of rank 2. It will therefore be sufficient to prove that $s \in G'$ when s is singular of index 2 in H_N. We can then find two linearly independent vectors y_1, y_2 of P and two linearly independent vectors x_1, x_2 of N such that $s \cdot y_1 = y_1 - x_2$, $s \cdot y_2 = y_2 + x_1$. The space X_0 spanned by x_1, x_2, y_1, y_2 is not isotropic of dimension 4; the conjugate R_0 of X_0 is therefore not isotropic and of dimension > 0. Its elements are left fixed by s, and it contains some nonsingular vector z. Let t_1, t_2, t_3, t_4 be the symmetries with respect to the conjugates of z, $z + x_1 + x_2$, $z + x_2$, $z + x_1$, and let $t = t_1 t_2 t_3 t_4$. Since K is of characteristic 2, we have

$$t_4 \cdot y_1 = y_1 + z + x_1 ,$$

$$t_3 t_4 \cdot y_1 = y_1 + z + x_1 ,$$

$$t_2 t_3 t_4 \cdot y_1 = y_1 + x_2 ,$$

$$t \cdot y_1 = y_1 + x_2 .$$

We see in the same way that $t \cdot y_2 = y_2 + x_1$. It is clear that t_i leaves x_1 and x_2 fixed ($i = 1, 2, 3, 4$) and that its restriction to R_0 is the same as that of t_1, which shows that t leaves the elements of R_0 fixed. Thus, we have $t = s$, which completes the proof.

It is easily verified that the case where K has 2 elements, $\dim M = 4$ and Q is of index 2 is actually exceptional. The group G' is then of index 2 in G.

1.6. Representation of G on the Multivectors

We make the same assumptions as in Section 5.

Let E be the exterior algebra over the space M. Then every operator $\sigma \in G$, being an automorphism of the vector space M, may be extended to an automorphism $\zeta(\sigma)$ of the algebra E; ζ is clearly a faithful representation of G by automorphisms of E. For any h, let E_h be the space of homogeneous elements of degree h of E. The operations $\zeta(\sigma)$, $\sigma \in G$, are homogeneous of degree 0; let $\zeta_h(\sigma)$ be the restriction of $\zeta(\sigma)$ to E_h. Then ζ_h is a representation of G, which is called the *representation on the h-vectors.*

Let also M^* be the dual of the space M. To any $\sigma \in G$ there is associated an automorphism ${}^t\sigma$ of M^*, the transpose of σ: if f is any linear form on M, then ${}^t\sigma \cdot f$ is the linear form $x \to f(\sigma \cdot x)$. Let $\sigma^* = {}^t\sigma^{-1}$; then $\sigma \to \sigma^*$ is a representation of G. Let E^* be the exterior algebra over M^*; then σ^* may be extended to an automorphism $\zeta^*(\sigma)$ of E^*, which is homogeneous of degree 0. Let $\zeta_h^*(\sigma)$ be the restriction of $\zeta^*(\sigma)$ to the space E^{*h} of homogeneous elements of degree h of E^*; then ζ_h^* is a representation of G, which is called the *representation on the h-covectors.*

The representations ζ_h, ζ_h^* are equivalent to each other. For, there is associated to B an isomorphism φ of M into M^* which assigns to every $x \in M$ the linear form $y \to B(x, y)$ on M. Let σ be in G; then, for $x, y \in M$, we have

$$\begin{aligned}(\sigma^* \cdot \varphi(x))(y) &= (\varphi(x))(\sigma^{-1} \cdot y) = B(x, \sigma^{-1} \cdot y)\\ &= B(\sigma x, y) = (\varphi(\sigma \cdot x))(y),\end{aligned}$$

since B is invariant under σ; thus, $\sigma^* = \varphi \circ \sigma \circ \varphi^{-1}$. The isomorphism φ may be extended to an isomorphism Φ of E with E^*; $\Phi \circ \zeta(\sigma) \circ \Phi^{-1}$ is an automorphism of E^* which extends σ^*, whence $\zeta^*(\sigma) = \Phi \circ \zeta(\sigma) \circ \Phi^{-1}$. It is clear that $\Phi(E^h) = E^{*h}$; if Φ_h is the restriction of Φ to E^h, then $\zeta_h^*(\sigma) = \Phi_h \circ \zeta_h(\sigma) \circ \Phi_h^{-1}$, which proves that ζ_h, ζ_h^* are equivalent to each other.

Let λ be any representation of a group Λ on a finite-dimensional vector space V; let V^* be the dual of V and, for any $\sigma \in \Lambda$, let ${}^t(\lambda(\sigma))$ be the transpose of $\lambda(\sigma)$, which is an automorphism of V^*. Let $\lambda^*(\sigma) = {}^t(\lambda(\sigma))^{-1}$; then λ^* is again a representation of Λ. Any representation μ of Λ which is equivalent to λ^* is said to be *contragredient* to λ. Let W be the space of μ. In order for λ, μ to be contragredient to each other, it is necessary and sufficient that there should exist a nondegenerate bilinear form β on $V \times W$ with the property that

$$\beta(\lambda(\sigma)\cdot x,\ \mu(\sigma)\cdot y) = \beta(x, y) \tag{1}$$

for all $\sigma \in \Lambda$, $x \in V$, $y \in W$.

For, assume first that λ and μ are contragredient to each other, and let φ be an isomorphism of W with V^* such that $\varphi(\mu(\sigma)\cdot y) = \lambda^*(\sigma)\cdot\varphi(y)$ for all $y \in W$. Then the bilinear form β: $(x, y) \rightarrow (\varphi(y))(x)$ satisfies condition (1), as can be verified immediately, and this bilinear form is nondegenerate because φ is an isomorphism. Conversely, assume that there exists a nondegenerate bilinear form β for which (1) is true. Then the mapping φ which assigns to every $y \in W$ the linear form $x \rightarrow \beta(x, y)$ is an isomorphism of W with V^* and we verify immediately that $\varphi(\mu(\sigma)\cdot y) = \lambda^*(\sigma)\cdot\varphi(y)$ for $\sigma \in \Lambda$, $y \in W$, which shows that μ is equivalent to λ^*.

If μ is contragredient to λ, then λ is to μ. Two representations which are both contragredient to a third one are equivalent to each other.

The representations ζ_h, ζ_h^* of G are not only equivalent but also contragredient to each other, as follows from the duality theory of exterior algebras.[1]

I.6.1. *Let G_1 be the group of operations of determinant* 1 *in G, and let ζ_h be the representation of G on the h-vectors* $(0 \le h \le m)$. *Then the representations of G_1 induced by ζ_h and ζ_{m-h} are equivalent to each other.*

Let e be a basic element of the one-dimensional space E^m. For any $\sigma \in G$, we have $\zeta(\sigma)\cdot e = (\det \sigma)e$, whence $\zeta(\sigma)\cdot e = e$ if $\sigma \in G_1$. If $u \in E^h$, $v \in E^{m-h}$, $u \wedge v$ is in E^m; set $u \wedge v = \beta(u, v)e$. Then β is a bilinear form on $E^h \times E^{m-h}$. It is well known that, for any $u \in E^h$, there is a $v \in E^{m-h}$ such that $u \wedge v = e$, which shows that β is nondegenerate. If $\sigma \in G_1$, $u \in E^h$, $v \in E^{m-h}$, then we have

$$\zeta(\sigma)\cdot u \wedge v = (\zeta_h(\sigma)\cdot u) \wedge (\zeta_{m-h}(\sigma)\cdot v).$$

Since $\zeta(\sigma)\cdot e = e$, we have

$$\beta(\zeta_h(\sigma)\cdot u,\ \zeta_{m-h}(\sigma)\cdot v) = \beta(u, v);$$

this shows that the representations of G_1 induced by ζ_h, ζ_{m-h} are contragredient to each other. Since ζ_h is contragredient to itself, I.6.1 is proved.

I.6.2. *Assume that the characteristic of K is $\neq$ 2. Then the representations ζ_h of G on the spaces of h-vectors* ($0 \le h \le m$, *where* $m = \dim M$) *are all simple, except in the following case: K has only* 3 *elements,* $m = 2$,

[1]N. Bourbaki, *op. cit.*, *Algèbre* III: (1947), Corollary to Proposition 1, Section 8, No. 2.

$h = 1$, and Q is of index 1. Let G^+ be the group of operations of determinant 1 in G and ζ^+_h the representation of G^+ induced by ζ_h. If $2h \neq m$, then ζ^+_h is simple. If $2h = m$, then ζ^+_h is either simple or equivalent to the sum of two simple representations; if ζ^+_h is not simple and if we are not considering the exceptional case mentioned above, then the two simple representations into which ζ^+_h splits are inequivalent to each other and also inequivalent to all ζ^+_k for $k \neq h$.

Since K is not of characteristic 2, M has a base $(x_1, \cdots, x_m)$ composed of mutually orthogonal vectors. For any subset A of $\{1, \cdots, m\}$ composed of h elements $i_1, \cdots i_h$, with $i_1 < \cdots < i_h$, set

$$\xi(A) = x_{i_1} \wedge \cdots \wedge x_{i_h} ;$$

the elements $\xi(A)$ form a base of E_h. Let H be the group of all automorphisms $s(\epsilon_1, \cdots, \epsilon_m)$ of M, where

$$s(\epsilon_1, \cdots, \epsilon_m) \cdot x_i = \epsilon_i x_i \qquad (1 \leq i \leq m),$$

the ϵ_i's being ± 1. It is clear that $H \subset G$ and that $H \cap G^+$ is composed of the $s(\epsilon_1, \cdots, \epsilon_m)$ for which

$$\prod_{i=1}^{m} \epsilon_i = 1.$$

We have

$$\zeta_h(s(\epsilon_1, \cdots, \epsilon_m)) \cdot \xi(A) = \chi_A(s(\epsilon_1, \cdots, \epsilon_m))\xi(A),$$

where

$$\chi_A(s(\epsilon_1, \cdots, \epsilon_m)) = \prod_{i \in A} \epsilon_i .$$

This shows that the representation $(\zeta_h)^H$ of H induced by ζ_h is equivalent to the sum of $C(m, h)$ representations of degree 1, say $\theta_{h,A}$. If A and A' are two distinct sets of h elements, then the functions χ_A, $\chi_{A'}$ are distinct; moreover, their restrictions to H^+ are distinct except in the case where $h = m/2$ and A, A' are complementary to each other. For, if i is in A but not in A', and if we set $\epsilon_i = -1$, $\epsilon_j = 1$ for $j \neq i$, then we have

$$\chi_A(s(\epsilon_1, \cdots, \epsilon_m)) \neq \chi_{A'}(s(\epsilon_1, \cdots, \epsilon_m),$$

which shows that $\chi_A \neq \chi_{A'}$. Except in the case where $h = m/2$ and A, A' are complementary to each other, it is easily seen that we can find an index k which either belongs to both A and A' or does not belong to either of them; i being selected as above, set $\epsilon_i = \epsilon_k = -1$, $\epsilon_j = 1$ for $k \neq i, j$; then $s(\epsilon_1, \cdots, \epsilon_m)$ is in H^+ and $\chi_A(s(\epsilon_1, \cdots, \epsilon_m)) \neq \chi_{A'}$

$(s(\epsilon_1, \cdots, \epsilon_m))$, which proves that the restrictions of χ_A, $\chi_{A'}$ to H^+ are distinct. Thus, we see that $(\zeta_h)^H$ splits into mutually inequivalent representations of H, and that the same is true of the representation $(\zeta_h^+)^{H^+}$ of H^+ induced by ζ_h if $h \neq m/2$. It follows that any subspace of E_h which is mapped into itself by the operations of $\zeta_h(H)$ is spanned by a certain number of the elements $\xi(A)$, and that the same is true if we assume only that the space is mapped into itself by the operations of $\zeta_h(H^+)$ and that $h \neq m/2$.

Now, let U be a subspace $\neq \{0\}$ of E_h which is mapped into itself by the operations of $\zeta_h(G^+)$; if $h = m/2$, assume further that U is mapped into itself by the operations of $\zeta_h(G)$. Then, for *any* base $(x'_1, \cdots, x'_m)$ of M composed of mutually orthogonal vectors, U has a base composed of elements of the form

$$x'_{i_1} \wedge \cdots \wedge x'_{i_h}.$$

Assume that $\xi(A) \in U$ for some $A = \{i_1, \cdots, i_h\}$, and suppose first that K has more than 3 elements. Let i be an index belonging to A and j an index not belonging to A. Then, for $a \in K$, we have $Q(x_i + ax_j) = Q(x_i) + a^2Q(x_j)$, and, since K has more than 3 elements, we may select $a \neq 0$ such that $Q(x_i + ax_j) \neq 0$. It is then clear that we can find a $b \neq 0$ in K such that $x_i + bx_j$ is orthogonal to $x_i + ax_j$. Let $x'_k = x_k$ if $k \neq i, j$, $x'_i = x_i + ax_j$, $x'_j = x_i + bx_j$; then $(x'_1, \cdots, x'_m)$ is a base of M composed of mutually orthogonal vectors, and $x_i = cx'_i + dx'_j$ with $c \neq 0$, $d \neq 0$; we have $\xi_A = c\xi' + d\xi''$, where

$$\xi' = x'_{i_1} \wedge \cdots \wedge x'_{i_h}$$

and ξ'' is the product derived from ξ' by replacing in it the factor x'_i by x'_j. From what we have said above, it follows that ξ', ξ'' are in U. Let B be the set obtained from A by replacing i by j; since x_j is a linear combination of x'_i, x'_j, $\xi(B)$ is a linear combination of ξ', ξ'', whence $\xi(B) \in U$. Thus, if $\xi(A) \in U$, then $\xi(B) \in U$ whenever B is obtained from A by replacing one of its elements by an index not occurring in it. It follows immediately that every $\xi(A)$ belongs to U, whence $U = E_h$. This proves that ζ_h is simple and that ζ^+_h is simple if $h \neq m/2$. Suppose now that K has 3 elements, and set $a_i = Q(x_i) = \pm 1$. If $a_i = a_j = -1$, then the space spanned by x_i, x_j is also spanned by $x_i + x_j$, $x_i - x_j$, which are orthogonal to each other, and $Q(x_i + x_j) = Q(x_i - x_j) = 1$. It follows that, by a suitable choice of the base $(x_i, \cdots, x_m)$, we may assume that at most 1 of the elements a_i is -1. Moreover, the same argument as above shows that, if $\xi(A) \in U$, and if $i \in A$, j is not in A, and $a_i = a_j$, then $\xi(B) \in A$, where B is the set deduced from A by replacing in it i by j. Thus, we have $U = E_h$ if all a_i are equal to $+1$. If not, let $a_i = +1$ for $i < m$, $a_m = -1$. Let $\mathfrak{a}_1$ be the set of those subsets A of $\{1, \cdots, m\}$ with h elements which contain m, and $\mathfrak{a}_2$ the set of those which do not. If $\xi(A) \in U$ for some $A \in \mathfrak{a}_i$, then the same is true for any other $A' \in \mathfrak{a}_i$ $(i = 1, 2)$. Assume further in this case that $m \geq 3$, and set $x'_{m-2} = x_{m-2} + x_{m-1}$, $x'_{m-1} = x_{m-2} - x_{m-1}$, $x'_k = x_k$ for $k \neq m-1, m-2$. Then $(x'_1, \cdots, x'_m)$ is a base of M composed of mutually orthogonal vectors and $Q(x'_i) = 1$ if $i < m-2$, $Q(x'_i) = -1$ if $i \geq m-2$. If $A = \{i_1, \cdots, i_h\}$, $i_1 < \cdots < i_h$, set

$$\xi'(A) = x'_{i_1} \wedge \cdots \wedge x'_{i_h} .$$

Assume that $m \in A$ implies $\xi(A) \in U$; then clearly, $m \in A$ also implies $\xi'(A) \in U$. If $h = m$, then $U = E_h$. If not, let A be a set containing m but not $m-1$, and let B be the set obtained by replacing m by $m-1$ in A. Since $Q(x'_{m-1}) = Q(x'_m)$ and $\xi'(A) \in U$, we have $\xi'(B) \in U$. But B does not contain m, and $\xi'(B)$ is a linear combination of the $\xi(B')$ for $B' \in \mathfrak{a}_2$. It follows that U must contain some $\xi(B')$ with $B' \in \mathfrak{a}_2$, and therefore that $U = E_h$. Similarly, if $A \in \mathfrak{a}_2$ implies $\xi(A) \in U$, then we have also $\xi'(A) \in U$ if $A \in \mathfrak{a}_2$. Let, then, A be a set of $\mathfrak{a}_2$ containing $m-1$ and B the set obtained by replacing $m-1$ by m in A; then $\xi'(B) \in U$, and it follows that $U = E_h$. If $m = 2$, $a_1 = 1$, $a_2 = -1$, then Q is of index 1, since $x_1 + x_2$ is singular. The cases $m = 0, 1$ being trivial, we see that ζ_h is always simple unless we are considering the exceptional case of the statement I.6.2 and that ζ_h^+ is simple if $h \neq m/2$. Assume now that $m = 2r$, $h = r$ and that $\zeta_h(s)$ maps U into itself for all $s \in G^+$. Disregarding the obvious case $m = 0$, there is an operation t in G but not in G^+, and G is the union of G^+ and G^+t. Since $t^2 \in G^+$, it is clear that $U + t(U)$ is mapped into itself by all operations of $\zeta_h(G)$. If we are not considering the exceptional case, this implies that $U + t(U) = E_r$. Assume further that U has been taken of the smallest possible dimension among the spaces $\neq \{0\}$, which are mapped into themselves by the operations of $\zeta_r(G^+)$. Since $G^+t = tG^+$, $t(U)$ is mapped into itself by the operations of $\zeta_h(G^+)$, and so is $U \cap t(U)$. The latter space is therefore either $\{0\}$ or U. If it is $\{0\}$, then E_r is the direct sum of U and $t(U)$; the representation of G^+ on the space U being simple by construction of U, the same is true of its representation on $t(U)$, and ζ_r^+ is equivalent to the sum of two simple representations. If $U \cap t(U) = U$, then we have $t(U) = U$ and $U = U + t(U) = E_r$, in which case ζ_r^+ is simple. In the exceptional case, we have $m = 2$, K has 3 elements, $Q(x_1) \neq Q(x_2)$, and the only nonsingular vectors are $\pm x_1$, $\pm x_2$. Since $Q(x_1) \neq Q(x_2)$, the group G is then identical to the group H introduced above, and Kx_1, Kx_2 are mapped into themselves by all operations of G. These spaces give two inequivalent representations of G, but two equivalent simple representations of G^+.

It remains to prove that, if ζ_r^+ is not simple and if we are not in the exceptional case, then the two simple representations of which ζ_r^+ is composed are inequivalent to each other and to all representations ζ_k^+, $k \neq r$. Let $\mathfrak{R}$ be the algebra of endomorphisms of E_r which commute with every $\zeta_r(s)$, $s \in G$, and $\mathfrak{R}^+$ the algebra of those which commute with every $\zeta_r(s)$, $s \in G^+$. Let σ be in $\mathfrak{R}$; then σ commutes in particular with the operations of $\zeta_r(H)$. Now we have seen that the representation $(\zeta_r)^H$ splits into mutually inequivalent representations, whose spaces are spanned by the elements $\xi(A)$. It follows that $\sigma \cdot \xi(A) = \lambda_A \xi(A)$, where λ_A is a scalar. For any scalar λ, let U_λ be the space spanned by those $\xi \in E_r$ such that $\sigma \cdot \xi = \lambda \xi$; since σ commutes with the operations of $\zeta_r(G)$, these operations map U_λ into itself, whence $U_\lambda = \{0\}$ or E_r, since ζ_r is simple. It follows that the λ_A's are all equal and that $\mathfrak{R} = K \cdot I$, where I is the identity mapping of E_r. Now, let $\mathfrak{a}$ and $\mathfrak{a}^+$ be the algebras of endomorphisms generated by $\zeta_r(G)$ and $\zeta_r(G^+)$, respectively. These algebras are semi-simple, since ζ_r, ζ^+_r are semi-simple. It follows that

$$[\mathfrak{a} : K \cdot I] \cdot [\mathfrak{R} : K \cdot I] = [\mathfrak{a}^+ : K \cdot I] \cdot [\mathfrak{R}^+ : K \cdot I] = (\dim E_r)^2.$$

On the other hand, let t be in G but not in G^+. Then we have $tG^+t^{-1} = G^+$, from which it follows that $\zeta_r(t)\ \mathfrak{a}^+\ \zeta_r(t^{-1}) = \mathfrak{a}^+$, and therefore that $\mathfrak{a}^+ + \zeta_r(t)\mathfrak{a}^+$ is an algebra. Since $G = tG^+$, this algebra contains $\zeta_r(G)$ and is therefore identical to $\mathfrak{a}$. We conclude that $[\mathfrak{a}:K \cdot I] = 2[\mathfrak{a}^+:K \cdot I]$, whence $[\mathfrak{R}^+: K \cdot I] = 2[\mathfrak{R}:K \cdot I] = 2$. Thus, $\mathfrak{R}^+$ is a commutative algebra of dimension 2. If $E_r = U + U'$, where U, U' are of dimension $\frac{1}{2} \dim E_r$ and mapped upon themselves by the operations of $\zeta_r(G^+)$, then the endomorphism τ which leaves the elements of U fixed but maps those of U' upon 0 is in $\mathfrak{R}^+$, and $\mathfrak{R}^+$ has zero divisors. Thus, $\mathfrak{R}^+$ is not simple, while it is well known that, were the representations of G^+ on U, U' equivalent to each other, then $\mathfrak{R}^+$ would be simple.

To every subset A of $\{1, \cdots, m\}$ we have associated above a homomorphism χ_A of the group H^+ into K. If A has r elements and A' has k elements, we have $\chi_A \neq \chi_{A'}$, for it is then always possible to find an index j, which is either in both A and A' or neither in A nor in A', and, proceeding as we did above, we can find an $s \in H^+$ such that $\chi_A(s) \neq \chi_{A'}(s)$. It follows that none of the representations of degree 1 of H^+ into which ζ_r^+ splits are equivalent to any of those into which ζ_k^+ splits. Therefore, the two simple representations of G^+ into which ζ_r^+ splits are inequivalent to all ζ_k^+, $k \neq r$, and I.6.2 is proved.

Remark. The proof of the fact that $\mathfrak{K} = K \cdot I$ does not make use of the fact that $h = r$. Thus, we see that, barring the exceptional case, the representations ζ_h are not only simple, but actually absolutely simple. We would see in the same way that ζ_h^+ is absolutely simple if $2h \neq m$; if $m = 2r$, $h = r$, and if ζ_r^+ splits into two simple representations, then these representations are absolutely simple.

We shall now determine under which condition ζ_r^+ is not simple. In order to do this, we shall construct a linear automorphism σ of the vector space (not the algebra!) E which commutes with all operations of $\zeta(G^+)$. If $x \in M$, then $y \rightarrow \frac{1}{2} B(x, y)$ is a linear function on M; it follows that there exists an antiderivation $\delta(x)$ of E such that $\delta(x) \cdot y = \frac{1}{2} B(x, y) \cdot 1$ for all $y \in M$. The operations $\delta(x)$ are homogeneous of degree -1, and $(\delta(x))^2 = 0$. Let $\mathfrak{E}$ be the algebra of endomorphisms of the vector space E. Since $(\delta(x))^2 = 0$, the linear mapping $x \rightarrow \delta(x)$ of M into $\mathfrak{E}$ may be extended to a homomorphism of E into $\mathfrak{E}$; we shall denote the image of a $\xi \in E$ under this homomorphism by $\delta(\xi)$. Since B is nondegenerate, $x \rightarrow \delta(x)$ is an isomorphism of the vector space M into $\mathfrak{E}$; it follows that $\xi \rightarrow \delta(\xi)$ is an isomorphism of E. Let e be a basic element of the one-dimensional space E_m ; set $\sigma(\xi) = \delta(\xi) \cdot e$. If $\xi = x_1 \wedge \cdots \wedge x_h$, $x_i \in M$, then $\delta(\xi) = \delta(x_1) \cdots \delta(x_h)$ is homogeneous of degree $-h$ (i.e., it maps E_k into E_{k-h} for any k); it follows that σ maps E_h into E_{m-h} . Let s be in G; then $\zeta(s)$ is an automorphism of E which maps each E_h onto itself. If $x \in M$, ξ, $\eta \in E$, we have $\delta(x) \cdot \xi \wedge \eta = (\delta(x) \cdot \xi) \wedge \eta + J(\xi) \wedge (\delta(x) \cdot \eta)$, where J is the main involution of E. Applying this formula to $\zeta(s) \cdot \xi$, $\zeta(s) \cdot \eta$ instead of ξ, η, and observing that $\zeta(s)$ commutes with J, we see immediately that $\zeta(s)\, \delta(x) (\zeta(s))^{-1}$ is an antiderivation. If $y \in M$, this antiderivation maps $\zeta(s) \cdot y = s \cdot y$ upon $\frac{1}{2} B(x, y) \cdot 1 = \frac{1}{2} B(s \cdot x, s \cdot y) \cdot 1$; it follows that $\zeta(s)\ \delta(x)\ (\zeta(s))^{-1} = \delta(s \cdot x)$. It follows immediately that $\zeta(s)\ \delta(\xi)\ (\zeta(s))^{-1} = \delta(\zeta(s) \cdot \xi)$ for any $\xi \in E$. Assume now that $s \in G^+$; then $\zeta(s) \cdot e = (\det s) e = e$, and we have $\sigma(\zeta(s) \cdot \xi) = \zeta(s) \cdot \sigma(\xi)$, which shows that σ commutes with $\zeta(s)$. Let us now determine the operation σ^2. Let $(x_1 , \cdots , x_m)$ be a base of M composed of mutually orthogonal vectors and assume that $e = x_1 \wedge \cdots \wedge x_m$; set $a_i = Q(x_i)$ and define the elements $\xi(A)$, for all subsets A of $\{1, \cdots , m\}$, as in the proof of I.6.2. We have $\delta(x_i) \cdot x_j = 0$ if $i \neq j$; $\delta(x_i) \cdot x_i = a_i \cdot 1$. An easy computation then gives

$$\sigma(\xi(A)) = (-1)^{\Sigma_{k \in A}(k-1)} \left(\prod_{k \in A} a_k\right) \xi(A'),$$

where A' is the complementary set of A. Let

$$D = \prod_{i=1}^{m} a_i ;$$

$2^{2r}D$ is the discriminant of B with respect to the base $(x_1, \cdots, x_m)$. We then have

$$\sigma^2(\xi) = (-1)^{\frac{1}{2}m(m+1)-m} D\xi = (-1)^{\frac{1}{2}m(m-1)} D\xi.$$

Assume now that $m = 2r$, and let σ_r be the restriction of σ to E_r, I_r the identity mapping of E_r. It is clear that σ_r, I_r are linearly independent and therefore form a base of the algebra denoted by $\mathfrak{R}^+$ in the proof of I.6.2. The representation ζ_r^+ splits or not according as to whether $\mathfrak{R}^+$ has zero divisors $\neq 0$ or not. Since $(-1)^{\frac{1}{2}m(m-1)} = (-1)^r$, we obtain:

I.6.3. *Let Δ be the discriminant of B with respect to some base of M. Assume that $m = 2r$. Then ζ_r^+ is simple if $(-1)^r \Delta$ is not a square in K and splits into two simple representations if $(-1)^r \Delta$ is a square in K.*

I.6.4. *If $m = 2r$ and Q is of maximal index r, then ζ_r^+ splits into two simple representations.*

For, in that case, M has a base $(x_1, \cdots, x_m)$ such that $B(x_i, x_j) = 1$ if $i = 2k - 1$, $j = 2k$ or $i = 2k$, $j = 2k - 1$ (where $1 \leq k \leq r$) and is 0 otherwise. The discriminant of B with respect to this base is $(-1)^r$, which proves I.6.4.

We propose now to investigate the representation ζ_h' of the commutator subgroup G' of G induced by ζ_h. In order to do this, we need some auxiliary results.

Let P be a nonisotropic plane (2-dimensional subspace) in M. An operation $s \in G^+$ which leaves all elements of the conjugate of P fixed is called a *plane rotation* of plane P.

I.6.5. *The field K being of characteristic $\neq 2$, the group G^+ of operations of determinant 1 in G is generated by the plane rotations.*

This is obvious if the dimension m of M is 1 or 2; assume $m > 2$. Let $\mathfrak{H}$ be the set of hyperplanes whose conjugates contain nonsingular vectors; if $H \in \mathfrak{H}$, let t_H be the symmetry with respect to H. Then G is generated by the operations t_H (I.5.1), and det $t_H = -1$, which shows that G^+ is generated by the products $t_H t_{H'}$, H, H' in $\mathfrak{H}$. If $H \cap H'$ is not isotropic, let P be its conjugate; then $t_H t_{H'}$ is a rotation of plane P. Assume now that $H \cap H'$ is isotropic. Let z be a vector $\neq 0$ in the intersection of $H \cap H'$ and its conjugate, and let z' be a singular vector in H such that $B(z, z') = 1$ (observe that, K not being of characteristic 2, H is not isotropic). Let P be the conjugate of $Kz + Kz'$ with respect to the restriction of B to $H \times H$. Then, since $Kz + Kz'$ is not isotropic, P is a nonisotropic subspace of dimension $m - 3$ of $H \cap H'$. Let N be its conjugate, which is of dimension 3; then $N \cap H$ and $N \cap H'$ (which are the conjugates of P with respect to the restrictions of B to $H \times H$ and $H' \times H'$) are nonisotropic subspaces of N; we have $N \cap H = Kz + Kz'$, $z \in N \cap H'$. Let x'_1 be a nonsingular vector in $N \cap H'$. If $B(x'_1, z') \neq 0$, set $x' = x'_1$; if not, let k be an element $\neq 0$ of K such that $Q(x'_1 + kz) = Q(x'_1) + kB(x'_1, z) \neq 0$ (there exists such an element, since $Q(x'_1) \neq 0$ and K has more than 2 elements); then set $x' = x'_1 + kz$, whence $B(x', z') = k \neq 0$. The element x' is a nonsingular element of $N \cap H'$, and $N \cap H' = Kz + Kx'$; since $N \cap H'$ is not isotropic, we have $B(z, x') \neq 0$. The conjugate of Kx' has a vector $x \neq 0$ in common with $N \cap H$. Since $B(x', z) \neq 0$, $B(x', z') \neq 0$, x is not in Kz or Kz'. But it is clear that the only singular elements of $N \cap H$ are those of $Kz \cup Kz'$; thus, x is not singular. Let $H'' = P + Kx + Kx'$. Since $Q(x) \neq 0$, $Q(x') \neq 0$, $B(x, x') = 0$, $Kx + Kx'$ is a nonisotropic subspace of the conjugate of P, and H'' is a nonisotropic hyperplane, whence $H'' \in \mathfrak{H}$. The spaces $H \cap H'' = P + Kx$, $H' \cap H'' = P + Kx'$ are not isotropic. Now, we may write $t_H t_{H'} = (t_H t_{H''})\,(t_{H''} t_{H'})$, and from what was said above, $t_H t_{H''}$ and $t_{H''} t_{H'}$ are plane rotations, which shows that $t_H t_{H'}$ is a product of two plane rotations; I.6.5 is thereby proved.

Consider now the case where $m = 3$. We shall establish that the representation ζ'_1 of G' is then simple. The notations $\mathfrak{H}$, t_H being as in the proof of I.6.5, we observe that, if $s \in G$, $H \in \mathfrak{H}$, then $t_{s(H)} t_H \in G'$, for it is clear that $t_{s(H)} = s t_H s^{-1}$. Were ζ'_1 not simple, then there would exist a one-dimensional subspace N of M which would be mapped into itself by the operations of $\zeta'_1(G')$. For, if N_1 is a 2-dimensional subspace of M which is invariant by the operations of $\zeta'_1(G')$, then so is the conjugate N of N_1. Assume for a moment that this is the case. Let x be a basic vector of N. If $Q(x) = 0$, let x' be a singular vector such that $B(x, x') = 1$, and $H = Kx + Kx'$, whence $H \in \mathfrak{H}$. Let x'' be an element $\neq 0$ of the conjugate of H, whence $Q(x'') \neq 0$. Then H contains a vector y such that $Q(y) = Q(x'')$ (by I.3.3). Let s be an operation of G such that $s \cdot x'' = y$. Then we have $t_H x = x$, but we see immediately that $t_{s(H)} x$ is not in $Kx = N$; thus, $t_{s(H)} t_H \cdot x$ is not in N, which results in a contradiction. If $Q(x) \neq 0$, let H_0 be the conjugate of Kx. Since ζ_1 induces a simple representation of G, there is an $s \in G$ such that $s(H_0) \neq H_0$. If x is not in $s(H_0)$, this nonisotropic plane contains at least one nonsingular vector not in $H_0 \cap s(H_0)$, which is of dimension 1 (as follows immediately from the fact that K has more than 2 elements). If K has more than 3 elements and $Kx \subset s(H_0)$, it is easily seen that $s(H_0)$ contains a nonsingular vector which is neither in Kx nor in H_0. In that case, H_0 contains a nonsingular vector y such that $s \cdot y$ is neither in Kx nor in H_0. Let H be the conjugate hyperplane of Ky; then $t_H \cdot x = x$, but $t_{s(H)} \cdot x$ is not in Kx, and we again have a contradiction. Assume now that K has only 3 elements. In that case, it is easily seen that Q takes all values $\neq 0$ (i.e., 1 and -1) in H_0. On the other hand, since s does not transform H_0 into itself and is a product of symmetries with respect to hyperplanes in $\mathfrak{H}$, there is at least one $H' \in \mathfrak{H}$ such that $t_{H'}(H_0) \neq H_0$, whence $t_{H'}(Kx) \neq Kx$. If y' is a vector $\neq 0$ in the conjugate of H', then there is a $y \in H_0$ such that $Q(y) = Q(y')$; thus, there is an $s' \in G$ such that $s'(y) = y'$. If H is the conjugate of Ky, then $H' = s'(H)$ and $t_{s'(H)}t_H$ does not transform Kx into itself. Thus, our assertion that ζ'_1 is simple if $m = 3$ is proved.

Still assuming that $m = 3$, let $\mathfrak{E}$ be the algebra of all endomorphisms of M and $\mathfrak{E}'$ the subalgebra of $\mathfrak{E}$ generated by G'. Since $\mathfrak{E}'$ admits a faithful simple representation of degree 3, $\mathfrak{E}'$ is a simple algebra. The dimension of $\mathfrak{E}'$ over its center is the square of a number which divides 3; thus, $\mathfrak{E}'$ is either $\mathfrak{E}$ or a commutative subfield of $\mathfrak{E}$. Now, let H be in $\mathfrak{H}$ and s an operation of G such that $s(H) \neq H$; then $t_{s(H)}t_H = s'$ is in G', is distinct from the identity I, and leaves invariant any vector z in $s(H) \cap H$. Since $s(H) \cap H$ is of dimension 1, $s' - I$, which is an element $\neq 0$ of $\mathfrak{E}'$, is not invertible (because $(s' - I) \cdot z = 0$); it follows that $\mathfrak{E}'$ is not a field, whence $\mathfrak{E}' = \mathfrak{E}$.

The space $\mathfrak{E}''$ spanned by the elements $s_1 - s_2$, s_1, $s_2 \in G'$, is obviously an ideal in $\mathfrak{E}'$, and $\mathfrak{E}' = \mathfrak{E}'' + KI$. Since $\mathfrak{E}'$ is simple, we have $\mathfrak{E}'' = \mathfrak{E}$. It follows that, if L is a linear function on $\mathfrak{E}$ which remains constant on G', then $L = 0$.

Assume now that m is ≥ 3, but otherwise arbitrary. If Z is any 3-dimensional nonisotropic subspace of M, we denote by H_Z the group of operations in G which leave invariant the elements of the conjugate of Z, by H_Z^+ the group $H_Z \cap G^+$, and by H_Z' the group $H_Z \cap G'$. The restrictions to Z of the operations of H_Z (respectively: H_Z^+, H_Z') include all operations of the orthogonal group of the restriction of Q to Z (respectively: all operations of determinant 1 in this group, all operations of the commutator subgroup of this group). We select a base in Z, and if $s \in H_Z$, we denote by $\sum(s)$ the matrix which represents the restriction of s to Z with respect to this base. Let θ be a linear representation of G^+ on a vector space T; assume that the following condition is satisfied: for any choice of Z in M (satisfying the conditions indicated above) and for any $u \in T$, the coefficients of the expression of $\theta(s) \cdot u$, where $s \in H_Z^+$, as a linear combination of the elements of a base of T may be expressed as polynomials of degrees ≤ 1 in the coefficients of $\sum(s)$. Let U be a subspace of T which is mapped into itself by all operations of $\theta(G')$; then we shall see that U is mapped into itself by all operations of $\theta(G^+)$. Making use of I.6.5 and observing that any nonisotropic plane is contained in some nonisotropic 3-dimensional space, we see that it will be sufficient to prove that U is mapped into itself by all operations of $\theta(H_Z^+)$ when Z is any nonisotropic 3-dimensional subspace of M. Let u be in U and let L be a linear function on T which vanishes on U. Then we may write $L(\theta(s)\cdot u) = L_1(s) + a_1$ for $s \in H_Z^+$, where L_1 is a linear form on $\mathfrak{E}$ and a_1 is a constant. We have $L(\theta(s)\cdot u) = 0$ if $s \in H'_Z$; thus, L_1 remains constant on H'_Z, whence $L_1 = 0$, as we have proved above. Since $L(u) = 0$, we have $a_1 = 0$, and $L(\theta(s)\cdot u) = 0$ for all $s \in H_Z^+$. This being true for any linear function on T which vanishes on U, we have $\theta(s)\cdot u \in U$, which proves our assertion.

We apply this to the case where $\theta = \zeta_h$, for some $h > 0$. We shall prove that the condition indicated above is satisfied. Let $(x_1, x_2, x_3, \cdots, x_m)$ be a base of M composed of mutually orthogonal vectors such that (x_1, x_2, x_3) is a base of Z. Let $\bar{E}$ be the subalgebra of E generated by $x_4, \cdots, x_m$ and $\bar{E}_k$ the space of homogeneous elements of degree k of $\bar{E}$; the elements of $\bar{E}$ are invariant by the operations of $\zeta_h(H_Z)$. If $\alpha \in \bar{E}_{k-3}$, let $A(\alpha)$ be the space spanned by $x_1 \wedge x_2 \wedge x_3 \wedge \alpha$; if $\alpha \in \bar{E}_{k-2}$, let $B(\alpha)$ be the space spanned by $x_1 \wedge x_2 \wedge \alpha, x_2 \wedge x_3 \wedge \alpha, x_3 \wedge x_1 \wedge \alpha$; if $\alpha \in \bar{E}_{k-1}$, let $C(\alpha)$ be the space spanned by $x_1 \wedge \alpha, x_2 \wedge \alpha, x_3 \wedge \alpha$; if $\alpha \in \bar{E}_h$, let $D(\alpha) = K\alpha$. Then E_h is the sum of the spaces $A(\alpha)$, $B(\alpha)$, $C(\alpha)$, $D(\alpha)$ (for all possible α) and the direct sum of some of these spaces. Identifying H_Z^+ to the group of operations of determinant 1 of the orthogonal group of the restriction of Q to Z, let ρ_k $(k = 0, 1, 2, 3)$ be the representation of this group on the k-vectors. Then we see that the representation of H_Z^+ induced by ζ_h is the sum of a certain number of representations each of which is equivalent to some ρ_k. But we know that ρ_0 and ρ_3 are trivial representations (they map every element of H_Z^+ upon the identity) and that ρ_1 is equivalent to ρ_2. It follows immediately that, for any $u \in E_h$, the coefficients of the expression of $\theta(s)\cdot u$ (for $s \in H_Z^+$) as linear combination of a base of E_h may be expressed as polynomials of degrees ≤ 1 in the coefficients of $\sum(s)$.

I.6.6. *Assume that M is of dimension ≥ 3 and that K is not of characteristic 2. Let G, G^+, and G' be the orthogonal group of Q, the group of operations of determinant 1 in G and the commutator subgroup of G. Let ζ_h be the representation of G on the h-vectors, and ζ_h^+, ζ'_h the representations of G^+, G' induced by ζ_h. If $2h \neq m$, then ζ'_h is simple. If $2h = m$ and ζ_h^+ is simple, then ζ'_h is simple. If $2h = m$ and ζ_h^+ is not simple, let $\zeta_{h,e}^+$ and $\zeta_{h,i}^+$ be the two simple representations of which it is the sum; then the* *representations* $\zeta'_{h,e}$, $\zeta'_{h,i}$ *of* G' *induced by* $\zeta_{h,e}^+$ *and* $\zeta_{h,i}^+$ *are simple. These representations are inequivalent to each other and to all* ζ'_k *for all* $k \neq h$.

Any subspace of E_h which is invariant by the operations of $\zeta'_h(G')$ is likewise invariant by those of $\zeta_h^+(G^+)$. It follows that ζ'_h is simple whenever ζ_h^+ is. If ζ_h^+ is not simple, then $\zeta_{h,e}^+$ and $\zeta_{h,i}^+$ are inequivalent to each other; the spaces T_e, T_i of these representations, together with $\{0\}$ and E_h, are therefore the *only* subspaces of E_h which are invariant by the operations of $\zeta^+_h(G^+)$; they are also the only ones invariant by the operations of $\zeta'_h(G')$, which shows that $\zeta'_{h,e}$, $\zeta'_{h,i}$ are inequivalent to each other. A similar argument, applied to the representation $\theta = \zeta_h + \zeta_k$ of G on $E_h + E_k (k \neq h)$, shows that $\zeta'_{h,e}$, $\zeta'_{h,i}$ are inequivalent to ζ'_k if $k \neq h$.

Consider now the case where $m = 2$. Assume that M contains a 1-dimensional space Kx which is invariant by all operations of G'. Suppose first that $Q(x) \neq 0$; then let H be the hyperplane Kx and t_H be the symmetry with respect to H. Let s be any operation in G and $t_{s(H)}$ the symmetry with respect to $s(H)$. Then $t_{s(H)}t_H$ transforms Kx into itself, whence $t_{s(H)} \cdot x \in Kx$, which shows that $s \cdot x$ is either in Kx or in its conjugate. If K has more than 3 elements or if Q is of index 0, there is an $s \in G$ such that $y = s \cdot x$ is not in Kx. Then we have $B(x, y) = 0$ and $Q(ax + by) = \alpha(a^2 + b^2)$ if $\alpha = Q(x)$; moreover, any vector z with $Q(z) = Q(x)$ is either in Kx or in Ky, which shows that $ab \neq 0$ implies $a^2 + b^2 \neq 1$. Setting $a = 2uv/(u^2 + v^2)$, $b = (u^2 - v^2)/(u^2 + v^2)$, we see that, if $u, v \neq 0$ in K, then v^2 is $\pm u^2$, which implies that K has 3 or 5 elements. Moreover, if K has 5 elements, then -1 is a square in K and Q is of index 1. It follows that, if Q is of index 0 and K has more than 3 elements, ζ'_1 is simple. If Q is of index 1, then $M = Kz + Kz'$, with singular vectors z, z', and it is easily seen that Kz, Kz' are mapped into themselves by all operations of G'; ζ'_1 is then not simple.

In the case where the basic field is of characteristic 2, it is easy to see that the representation of G on the h-vectors is in general not simple (not even semi-simple) if $h > 1$. In the case where $h = 1$, we have the following results:

I.6.7. *Assume that* K *is of characteristic* 2. *The representation of* G *on the space* M *is then simple except in the following case:* K *has* 2 *elements,* dim $M = 2$, *and* Q *is of index* 1. *The representation of the commutator subgroup* G' *of* G *on* M *is simple except in the following cases:* (*a*) dim $M = 2$ *and* Q *is of index* 1; (*b*) K *has* 2 *elements,* dim $M = 4$, *and* Q *is of index* 2.

Let N be a subspace of M distinct from $\{0\}$ and M which is mapped into itself by every operation of G' ; then the conjugate N' of N is likewise mapped into itself by every operation of G'.

We shall first discuss the case where N contains some nonsingular vector x. Let s be any element of G and $y = s \cdot x$; denote by t_x and t_y the symmetries with respect to the conjugates of Kx and Ky. Then $t_y = st_x s^{-1}$, whence $t_y t_x = t_y t_x^{-1} \in G'$, and therefore $t_y t_x(N) = N$. Since $x \in N$, we have $t_x(N) = N$, whence $t_y(N) = N$. If $z \in N$, then $t_y \cdot z = z - (Q(y))^{-1} B(z, y)y$. If, for some $z \in N$, we have $B(z, y) \neq 0$, then y is in N (since z and $t_y \cdot z$ are in N); if not, then y is in N'. Let U be the space spanned by all vectors $s \cdot x$, $s \in G$; then it is clear that U is mapped into itself by every operation of G, and it follows from what we have just said that $U \subset N + N'$. Let U' be the conjugate of U, and let v be any nonsingular vector in M; then the symmetry t_v with respect to the conjugate of Kv maps U into itself, from which it follows in the same manner as above that v lies either in U or in U'. Let $V = U + U'$, and let W be a subspace of M supplementary to V.

Assume first that $W \neq \{0\}$. Let w be an element $\neq 0$ in W, and v any element in V. Since all nonsingular elements of M are in V, we have $Q(w) = 0$, and $0 = Q(v + w) = Q(v) + B(v, w)$, or $Q(v) = B(v, w)$. The restriction of Q to V is therefore linear; in particular, if $k \in K$, $k^2 Q(v) = Q(kv) = kQ(v)$; taking v such that $Q(v) \neq 0$, we see that $k^2 = k$; i.e., K has only 2 elements. If z is any singular element $\neq 0$ of M, there is an $s \in G$ such that $s \cdot z = w$ (I.4.1); since $V = U + U'$ is mapped into itself by the operations of G, z cannot be in V, which shows that $Q(v) = B(v, w) \neq 0$ for all $v \neq 0$ in V. Were V of dimension > 1, it would contain at least one vector $v \neq 0$ such that $B(v, w) = 0$, which is impossible. Thus, $\dim V = 1$, whence $\dim U = \dim U' = 1$, and, since $\dim U' = \dim M - \dim U$, we have $\dim M = 2$. Since $Q(w) = 0$, Q is of index 1.

Assume now that $W = \{0\}$, whence $U + U' = M$. We shall see that $U' = \{0\}$ in that case. For, assume for a moment that U' contains an element $x' \neq 0$. Taking x to be $\neq 0$ in U, $x + x'$ is neither in U nor in U', whence $Q(x + x') = 0$. Since $B(x, x') = 0$, we have $Q(x) + Q(x') = 0$, $Q(x) = Q(x')$. But this implies the existence of an $s \in G$ such that $s \cdot x = x'$, in contradiction to the assumption that $s(U) = U$. Thus, we have $U' = \{0\}$, whence $U = N + N' = M$. Since $\dim N' = \dim M - \dim N$, it follows that $N \cap N' = \{0\}$. Since $B(x, x) = 0$ for every x, this implies $\dim N > 1$. Let x again be nonsingular in N, and let x' be $\neq 0$ in N'. Since $N \cap N' = \{0\}$, the restriction of B to $N \times N$ is nondegenerate, and N contains a vector y such that $b = B(x, y) \neq 0$.

We assert that $Q(x') = Q(y)$. Were this not the case, if we set $a = b(Q(y) + Q(x'))^{-1}$, then

$$Q(x + a(y + x')) = Q(x) + a^2Q(y) + ab + a^2Q(x') = Q(x),$$

and there would exist an $s \in G$ such that $s \cdot x = x + a(y + x')$. But a would be $\neq 0$, and $s \cdot x$ would lie neither in N nor in N', which is impossible. Thus, Q is constant on the set of elements $\neq 0$ in N'. This constant is $\neq 0$, while, otherwise, B would be 0 on $N' \times N'$ and N' would be contained in its conjugate N. Since $Q(kx') = Q(x')$ for every $k \in K$, $k \neq 0$, we have $k^2 = 1$, and K has only 2 elements. Let x' and x'' be linearly independent elements in N'; then we have $1 = Q(x' + x'') = Q(x') + Q(x'') + B(x', x'') = B(x', x'')$; were N' of dimension > 2, then it would clearly contain two linearly independent vectors orthogonal to each other, which is not the case. Thus, $\dim N = \dim N' = 2$, and $\dim M = 4$. Exchanging the roles played by N and N', we see that $Q(x) = 1$ for all $x \neq 0$ in N and $B(x, y) = 1$ if x, y are linearly independent in N. Let (x_1, x_2) be a base of N and (x'_1, x'_2) a base of N'. Set $z_1 = x_1 + x'_1$, $z_2 = x_2 + x'_2$; then we have $Q(z_1) = Q(z_2) = B(z_1, z_2) = 0$, and Q is of index 2.

Assume now that N is totally singular. Were N' not totally singular, we could replace N by N' in the preceding argument. Assume now that N and N' are totally singular, and let $r = \dim N$. Then N and N' are totally isotropic, and each one is contained in the other, whence $N = N'$. Since $\dim N' = \dim M - r$, we have $\dim M = 2r$, and M is the direct sum of N and of a totally singular space P. Let N_1 be any subspace of N and P_1 the intersection of P with the conjugate of N_1; then $N_2 = N_1 + P_1$ is totally singular of dimension r, and there exists an $s \in G$ such that $s(N) = N_2$ (I.4.1). Since G' is a normal subgroup of G, it is clear that N_2 is still mapped into itself by every operation of G'; the same is therefore true of $N_1 = N \cap N_2$ and of the conjugate N'_1 of N_1. If $r > 1$, then we may take N_1 to be $\neq \{0\}$ and N; then N'_1 is $\neq \{0\}$, M and is of dimension $> r$, which implies that it contains a nonsingular vector, and we are reduced to the previous case. If $r = 1$, then $\dim M = 2$ and Q is of index 1.

If $\dim M = 2$ and Q is of index 1, then we have $M = Kx + Ky$, with $Q(x) = Q(y) = 0$, $B(x, y) = 1$. The only singular vectors of M are those of $Kx \cup Ky$, which shows that the operations of G permute Kx and Ky among themselves. This permutation gives rise to a representation of G on the group of permutations of the set $\{Kx, Ky\}$, which is abelian; the kernel of this representation contains G', which shows that the operations of G' map Kx and Ky upon themselves. The automorphism of M which exchanges x and y is obviously in G, and neither of the spaces Kx, Ky is mapped into itself by the operations of G. If K has more than 2 elements, then no non totally singular subspace of M can be mapped into itself by every operation of G', and, a fortiori, of G. If K has only 2 elements, then M has only one non totally singular space of dimension 1, namely, $K(x + y)$, and this space is mapped into itself by every operation of G. Assume now that K has 2 elements, that $\dim M = 4$, and that Q is of index 2. Then M has a base (x_1, x_2, y_1, y_2) composed of singular vectors such that $B(x_i, x_j) = B(x_i, y_j) = B(y_i, y_j) = 0$ if $i \neq j$, $B(x_i, y_i) = 1$. Set $u = x_1 + y_1$, $v = x_1 + x_2 + y_2$, $N = Ku + Kv$; then the restriction of Q to N is of index 0, the conjugate N' of N is spanned by $u' = x_2 + y_2$, $v' = x_2 + x_1 + y_1$, and the restriction of Q to N' is of index 0. If $z \in N$, $z' \in N'$, $z \neq 0$, $z' \neq 0$, then we have $Q(z + z') = 1 + 1 = 0$; thus, every nonsingular vector is either in N or in N'. Conversely, let P be a 2-dimensional subspace which contains no singular vector $\neq 0$. Then we have $P \subset N \cup N'$, from which it follows easily that P is either N or N'. Thus, the operations of G permute N and N' among themselves, and, by the same argument as above, those of G' map both N and N' upon themselves. Moreover, it follows from our analysis that N and N' are the only subspaces $\neq \{0\}$, M which are mapped into themselves by all operations of G'. Since $Q(u) = Q(u')$, there is an $s \in G$ which maps u upon u', whence $s(N) = N'$; this shows that the representation of G on M is then simple.

CHAPTER II

THE CLIFFORD ALGEBRA

In this chapter, Q will denote a quadratic form on a vector space M of finite dimension m over a field K; B will denote the associated bilinear form of Q.

2.1. Definition of the Clifford Algebra

Let T be the tensor algebra of the vector space M, and I the ideal generated in T by the elements $x \otimes x - Q(x)\cdot 1$ for all $x \in M$. Then the factor algebra $C = T/I$ is called the *Clifford algebra* of the quadratic form Q.

The algebra T is a graded algebra; let T^h be the space of homogeneous elements of degree h of T. Denote by T_+ the sum of all spaces T^h for h even, by T_- the sum of the spaces T^h for h odd; T is then the direct sum of T_+ and T_-, and we have

$$T_+T_+ \subset T_+ \;; \qquad T_+T_- \subset T_- \;; \qquad T_-T_+ \subset T_- \;; \qquad T_-T_- \subset T_+ \,.$$

The ideal I is generated by elements belonging to T_+. Since T has a base composed of homogeneous elements, it is clear that every element of I may be written as a sum of elements of $I \cap T_+$ and $I \cap T_-$. Let C_+ and C_- be the vector spaces $T_+/(I \cap T_+)$ and $T_-/(I \cap T_-)$. Then, clearly, C is the direct sum of C_+ and C_- and

$$C_+C_+ \subset C_+ \;; \qquad C_+C_- \subset C_- \;; \qquad C_-C_+ \subset C_- \;; \qquad C_-C_- \subset C_+$$

The elements of C_+ are called *even*, those of C_- *odd*; the even elements form a subalgebra of C. The linear mapping J of C onto itself defined by $J(u) = u$ if $u \in C_+$, $J(u) = -u$ if $u \in C_-$ is an automorphism of C, called the *main involution*. If K is of characteristic 2, J is the identity.

Now, let h be any integer ≥ 0. The mapping $(x_1, \cdots, x_h) \to x_h \otimes \cdots \otimes x_1$ of M^h (the product of h times M by itself) into T^h is clearly multilinear. It follows that there exists a linear mapping α_h^T of T^h into itself such that $\alpha_h^T(x_1 \otimes \cdots \otimes x_h) = x_h \otimes \cdots \otimes x_1$ whenever $x_1, \cdots, x_h$ are in M. Let α^T be the linear mapping of T onto itself which extends all the mappings $\alpha^T{}_h$. It is clear that $(\alpha^T)^2$ is the identity. If $x_1, \cdots, x_h, y_1, \cdots, y_k$ are in M, $t = x_1 \otimes \cdots \otimes x_h$, $t' = y_1 \otimes \cdots \otimes y_k$, then we have

$$\alpha^T(t \otimes t') = y_k \otimes \cdots \otimes y_1 \otimes x_h \otimes \cdots \otimes x_1 = \alpha^T(t') \otimes \alpha^T(t),$$

which proves that α^T is an antiautomorphism of T. This antiautomorphism leaves the elements of $T^0 + T^1$ fixed; it maps upon themselves the generators $x \otimes x - Q(x)\cdot 1$ of I. Now, I is the set of all elements which are sums of products of the form $t \otimes (x \otimes x - Q(x)\cdot 1) \otimes t'$, with $t, t' \in T$, $x \in M$; it follows immediately that $\alpha^T(I) = I$. Thus, α^T defines in a natural manner a linear mapping α of $C = T/I$ onto itself. It is clear that α is an antiautomorphism of C, whose square is the identity; it is called the *main antiautomorphism* of C.

Let C' be an algebra over K. Assume that we have a linear mapping φ of M into a subspace M' of C' such that $(\varphi(x))^2 = Q(x)\cdot 1$ for all $x \in M$. Let π be the natural mapping of T onto $C = T/I$. Then there is a homomorphism ψ of C into C' such that $\psi(\pi(x)) = \varphi(x)$ for $x \in M$. For, we know that φ may be extended to a homomorphism Φ of T into C'. If $x \in M$, then $\Phi(x \otimes x - Q(x)\cdot 1) = (\varphi(x))^2 - Q(x)\cdot 1 = 0$; this shows that the kernel of Φ contains I. Thus, Φ may be factored in the form $\Phi = \psi \circ \pi$, where ψ is a homomorphism of C into C' with the required property. If M' generates C', then, clearly, we have $\psi(C) = C'$.

We shall now construct such an algebra C'. We start with the exterior algebra E on M, in which the multiplication will be denoted by the sign $\wedge$. We know that, λ being any linear function on M, there exists an antiderivation δ of E such that $\delta x = \lambda(x)\cdot 1$ for $x \in M$; δ is homogeneous of degree -1, and $\delta^2 = 0$. There exists a bilinear form B_0 on $M \times M$ such that $B_0(x, x) = Q(x)$ for all $x \in M$ (I.2.2). We denote by δ_x the antiderivation of E such that

$$\delta_x \cdot y = B_0(x, y)\cdot 1 \qquad (y \in M).$$

Let L_x be the operator $u \to x \wedge u$ of left multiplication by x in E; set $L'_x = L_x + \delta_x$. Then $x \to L'_x$ is a linear mapping φ of M into the algebra $\mathfrak{E}$ of endomorphisms of the vector space E. If $x \in M$, we have $\delta_x^2 = 0$ and

$$L_x\delta_x + \delta_x L_x = Q(x)\cdot I,$$

where I is the identity mapping. For, if $u \in E$, we have $\delta_x L_x \cdot u = \delta_x(x \wedge u) = (\delta_x x) \wedge u - x \wedge (\delta_x u) = Q(x)u - L_x\delta_x \cdot u$. Since $x \wedge x = 0$, we have $L_x^2 = 0$. It follows that $L'^2_x = Q(x)I$. This shows that there is a homomorphism ψ of C into $\mathfrak{E}$ such that $\psi(\pi(x)) = L'_x$ for $x \in M$. If

$x \in M$, then $L'_x \cdot 1 = x$, since $\delta_x \cdot 1 = 0$, and $x \to L'_x$ is an isomorphism of M. It follows immediately that π induces an isomorphism of M into C. We shall henceforth identify the elements of M with their images in C under M. Thus, M will be considered as a subspace of C. We have

$$x^2 = Q(x) \cdot 1 \qquad \text{if} \qquad x \in M. \tag{1}$$

Applying this to x, y, and $x + y$ (where $x, y \in M$) and remembering that $Q(x + y) - Q(x) - Q(y) = B(x, y)$, we obtain

$$xy + yx = B(x, y) \cdot 1 \qquad (x, y \in M). \tag{2}$$

The result established above becomes

II.1.1. *Let φ be a linear mapping of M into an algebra C' over K. Assume that $(\varphi(x))^2 = Q(x) \cdot 1$ for $x \in M$. Then φ may be extended to a homomorphism ψ of C into C'. If $\varphi(M)$ generates C', then $\varphi(C) = C'$.*

Let us now return to the homomorphism ψ of C into $\mathfrak{E}$ considered above. Set $\theta(u) = \psi(u) \cdot 1$ for $u \in C$. Then θ is a linear mapping of C into E. We remind the reader that an element $\Lambda \in \mathfrak{E}$ is called homogeneous of degree d if Λ transforms any homogeneous element of degree h of E into a homogeneous element of degree $h + d$. If $\Lambda_1, \cdots, \Lambda_k$ are homogeneous of respective degrees $d_1, \cdots, d_k$, then $\Lambda_1 \cdots \Lambda_k$ is homogeneous of degree $d_1 + \cdots + d_k$. For any $x \in M$, L'_x is homogeneous of degree $+1$ and δ_x of degree -1. Let $x_1, \cdots, x_h$ be in M. Then we have

$$\psi(x_1 \cdots x_h) = (L_{x_1} + i_{x_1}) \cdots (L_{x_h} + i_{x_h}),$$

and this may be written as

$$\psi(x_1 \cdots x_h) = L_{x_1} \cdots L_{x_h} + \sum_{d=-h}^{h-1} \Lambda_d, \tag{3}$$

where Λ_d is homogeneous of degree d. It follows that

$$\theta(x_1 \cdots x_h) = x_1 \wedge \cdots \wedge x_h + \sum_{d=0}^{h-1} \xi_d, \tag{4}$$

where ξ_d is homogeneous of degree d. For any h, let E^h be the space of homogeneous elements of degree h of E and $F_h = \sum_{d<h} E^d$. The space E^h is spanned by the products of h elements of M. Thus, it follows from (4) that $E_h \subset \theta(C) + F_h$. We have $F_0 = \{0\}$. It follows immediately by induction on h that $E_h \subset \theta(C)$ for every h, whence $\theta(C) = E$. We conclude that C is of dimension at least equal to the dimension 2^m of E.

II.1.2. *Let $(x_1, \cdots, x_m)$ be a base of M. If $\sigma = (i_1, \cdots, i_h)$ is a strictly increasing sequence of integers $i_1, \cdots, i_h$ between 1 and m, let $P(\sigma)$ be the product $x_{i_1} \cdots x_{i_h}$ in C. Then the elements $P(\sigma)$ form a base of C, which is of dimension 2^m.*

(Observe that, among the sequences σ, we include the empty sequence σ_0; $P(\sigma_0)$ is 1.) For any $h \geq 0$, let C_h be the space spanned by the $P(\sigma)$ for the sequences σ of length h, and set $D_h = \sum_{h' \leq h} C_{h'}$. It is clear that $C_0 = D_0 = K \cdot 1$, $C_1 = M$. We shall prove that, if $\sigma = (i_1, \cdots, i_h)$, then $x_i P(\sigma)$ is a linear combination of the elements $P(\sigma')$, where σ' runs over the sequences $(j_1, \cdots, j_{h'})$ such that $h' \leq h + 1$ and $j_1 \geq \min \{i, i_1, \cdots, i_h\}$. This is true if $h = 0$. Assume that it is true for $h - 1$, h being > 0. If $i < i_1$, then we have $x_i \cdot P(\sigma) = P(\sigma')$ with $\sigma' = (i, i_1, \cdots, i_h)$. If $i = i_1$, then we have $x_i P(\sigma) = Q(x_i)P((i_2, \cdots, i_h))$ by formula (1) above. If $i > i_1$, let $\sigma_1 = (i_2, \cdots, i_h)$; then it follows from formula (2) above that

$$x_i P(\sigma) = B(x_i, x_{i_1})P(\sigma_1) - x_{i_1} x_i P(\sigma_1),$$

and it follows from our inductive assumption that $x_i P(\sigma_1)$ is a linear combination of the $P(\sigma'')$ for the sequences $\sigma'' = (k_1, \cdots, k_{h''})$ such that $h'' \leq h$, $k_1 \geq \min \{i, i_2, \cdots, i_h\} > i_1$. For any such sequence, $\sigma' = (i_1, k_1, \cdots, k_{h''})$ is strictly increasing and

$$x_{i_1} P(\sigma'') = P(\sigma'),$$

and this proves our assertion for h. It follows that $x_i D_m \subset D_m$ for all i, whence $x D_m \subset D_m$ for all $x \in M$. Since M generates C, we have $v D_m \subset D_m$ for all $v \in C$, whence $v = v \cdot 1 \in D_m$ and $D_m = C$. There are exactly 2^m sequences σ; thus, D_m is of dimension $\leq 2^m$. But we know already that D_m is of dimension $\geq 2^m$. It follows that the elements $P(\sigma)$, which generate the vector space D_m, are linearly independent, which proves II.1.2.

II.1.3. *Let the notation be as in* II.1.1. *If $\varphi(M)$ generates C' and C' is of dimension $\geq 2^m$, then ψ is an isomorphism of C with C'.*

For we have $\psi(C) = C'$ and C is of dimension 2^m.

II.1.4. *Let N be a subspace of M. Then the subalgebra of C generated by N is isomorphic to the Clifford algebra of the restriction of Q to N.*

Let $(x_1, \cdots, x_m)$ be a base of M containing a base $(x_1, \cdots, x_n)$ of N. The products $x_{i_1} \cdots x_{i_h}$, where $i_1 < \cdots < i_h \leq n$ are linearly independent in the algebra D generated by N, whence $\dim D \geq 2^n$; II.1.4. then follows from II.1.3.

II.1.5. *Let K' be an overfield of K, $M^{K'}$ and $C^{K'}$ the vector space and the algebra deduced from M and C, respectively, by extension to K' of the basic field, and Q' the quadratic form on $M^{K'}$ which extends Q. Then $C^{K'}$ is isomorphic to the Clifford algebra of Q'.*

Let $(x_1, \cdots, x_m)$ be a base of M and $a_1, \cdots, a_m$ elements of K'. We have

$$\left(\sum_{i=1}^{m} a_i x_i\right)^2 = \sum a_i^2 x_i^2 + \sum_{i<j} a_i a_j (x_i x_j + x_j x_i)$$

$$= \sum_{i=1}^{m} a_i^2 Q(x_i) + \sum_{i<j} a_i a_j B(x_i, x_j) = Q'\left(\sum_{i=1}^{m} a_i x_i\right).$$

Since $C^{K'}$ is of dimension 2^m, II.1.5 follows from II.1.3.

Let us now return to the consideration of the linear mapping θ of C onto E introduced above. Since $\dim C = 2^m = \dim E$, θ is a linear isomorphism which coincides with the identity on $K \cdot 1$ and on M. We shall generally identify the underlying vector space of C with that of E by means of θ. If u, v are in E, uv will denote their product in C, while $u \wedge v$ will denote their product in E. It should be kept in mind, however, that our identification depends on the choice of a bilinear form B_0 such that $B_0(x, x) = Q(x)$.

We observe that, in formula (3) above, Λ_d can only be $\neq 0$ if $d \equiv h \pmod 2$, because a product of r operators L_{x_i} and $h - r$ operators δ_{x_i} is of degree $h - 2r$. On the other hand, C_+ (respectively: C_-) is obviously spanned by the products of an even (respectively: odd) number of factors in M. It follows that C_+ (respectively: C_-) is the set of elements of E whose homogeneous components $\neq 0$ are all of even (respectively: odd) degree. This shows that the main involution of C is the same as that of E. On the other hand, we see that

$$x_1 \cdots x_h \equiv x_1 \wedge \cdots \wedge x_h \qquad (\text{mod} \sum_{h' \leq h-2} E_{h'})$$

for any $x_1, \cdots, x_h$ in M. We have therefore obtained the following results:

II.1.6. *Let there be given a bilinear form B_0 on $M \times M$ such that $Q(x) = B_0(x, x)$ for $x \in M$. We can then identify the underlying vector space of the Clifford algebra C with that of the exterior algebra E of M in such a way that, for any $x \in M$, the operator of left multiplication by x in C is $L_x + \delta_x$, where L_x is the operator of left multiplication by x in E and δ_x the antiderivation of E such that $\delta_x \cdot y = B_0(x, y) \cdot 1$ for $y \in M$. Let C_h be the subspace of C spanned by the products of at most h elements of*

M and E_h the space of homogeneous elements of degree h of E; then we have $C_h = \sum_{h' \leq h} E_{h'}$. If $x_1, \cdots, x_h$ are in M, then we have, for $h \geq 2$, $x_1 \cdots x_h \equiv x_1 \wedge \cdots \wedge x_h \pmod{C_{h-2}}$. We have $C_+ = \sum_{h \text{ even}} E_h$, $C_- = \sum_{h \text{ odd}} E_h$.

2.2. Structure of the Clifford Algebra

II.2.1. *Assume that M is of even dimension $2r$ and that Q is of rank m and defect 0. Then the Clifford algebra C of Q is a central simple algebra. If Q is furthermore of index r, then C is isomorphic to the algebra of all matrices of degree 2^r with coefficients in K.*

Let K' be an algebraically closed overfield of K, $M^{K'}$ the vector space deduced from M by extension to K' of the basic field, and Q' the quadratic form on $M^{K'}$ which extends Q. Then Q' is still of rank m and defect 0, and is of index r (by I.3.4). Taking II.1.5. into account, we see that it suffices to prove II.2.1 in the case where Q is of index r. Assume that this is the case. Let N and P be two totally singular subspaces of M which are supplementary to each other and of dimension r. Let $(x_1, \cdots, x_r)$ and $(y_1, \cdots, y_r)$ be bases of N and P such that $B(x_i, y_j) = \delta_{ij}$ $(1 \leq i, j \leq r)$. Let B_0 be the bilinear form on $M \times M$ defined by the conditions

$$B_0(x_i, x_j) = B_0(y_i, y_j) = B_0(x_i, y_j) = 0;$$

$$B_0(y_i, x_j) = \delta_{ij} \qquad (1 \leq i, j \leq r)$$

It is easily seen that $B_0(x, x) = Q(x)$ for all $x \in M$. The form B_0 vanishes on $N \times N$, on $N \times P$, and on $P \times P$, and its restriction to $P \times N$ is nondegenerate. Using B_0, we identify the space C to the underlying vector space of the exterior algebra E on M, as explained in II.1.6. Let E^N and E^P be the subalgebras of E generated by N and P, respectively. We use the same notation as in Section 1. If $x \in N$, then $\delta_x(N) = \{0\}$; since δ_x is an antiderivation, $\delta_x(E^N) = \{0\}$ and it follows that $xu = x \wedge u$ for all $u \in E^N$. This shows that E^N is identical (as an algebra) with the subalgebra of C generated by N. We see in the same way that E^P is a subalgebra of C. We set $f = y_1 \cdots y_r = y_1 \wedge \cdots \wedge y_r$, and we consider the left ideal Cf of C. The elements $x_{i_1} \cdots x_{i_p} y_{j_1} \cdots y_{j_q}$, where $i_1 < \cdots < i_p \leq r$, $j_1 < \cdots < j_q \leq r$, form a base of C (by II.1.2); and we have $y_i f = y_i \wedge f = 0$ $(1 \leq i \leq r)$. It follows immediately that the elements $x_{i_1} \cdots x_{i_p} f$ form a base of Cf, i.e., that $u \rightarrow uf$ $(u \in E_N)$ is a linear isomorphism of E_N with Cf; Cf is therefore of dimension 2^r. To every element $w \in C$ we may associate the endomorphism $\rho(w)$ of E_N defined by the condition that

$$wuf = (\rho(w) \cdot u)f$$

for all $u \in E^N$. It is clear that ρ is a linear representation. If $w \in E_N$, it is clear that $\rho(w)$ is the operator of left multiplication by w in E_N. Since B_0 vanishes on $N \times P$, we see that, if $x \in N$, δ_x maps P, and therefore also E^P, upon $\{0\}$. It follows that $xv = x \wedge v$ if $v \in E^P$, whence $uf = u \wedge f$ for all $u \in E^N$. Now, let y be in P. Then we have $yuf = y \wedge uf + \delta_y(uf)$. The first term is $y \wedge (u \wedge f) = 0$, since y divides f. We have $\delta_y(uf) = (\delta_y u) \wedge f + J(u) \wedge \delta_y f$, but $\delta_y f = 0$, since B_0 vanishes on $P \times P$, whence $yuf = (\delta_y u)f$. Since δ_y maps N into $K \cdot 1$, it maps E^N into itself. It follows that

$$\rho(y) \cdot u = \delta_y \cdot u \qquad (y \in P, u \in E^N).$$

The operation δ_{y_i} maps x_i upon 1, x_j upon 0 if $j \neq i$. Let $e = x_r \cdots x_1$. Since δ_{y_i} is an antiderivation, we see easily that $\delta_{y_{h+1}} \cdots \delta_{y_r}$ maps $x_r \cdots x_{h+1}$ upon 1 for any h, whence $\rho(f) \cdot e = 1$. On the other hand, $\rho(f)$, which is homogeneous of degree $-r$, maps any homogeneous element of degree $< r$ of E^N upon 0. Let $\sum$ be the set of strictly increasing sequences of integers between 1 and r; if $\sigma = (i_1, \cdots, i_h)$, set

$$\xi(\sigma) = x_{i_1} \cdots x_{i_h}.$$

We shall see that, given σ and σ_1 in $\sum$, there is a $w \in C$ such that $\rho(w) \cdot \xi(\sigma) = \xi(\sigma_1)$, $\rho(w) \cdot \xi(\sigma') = 0$ if $\sigma' \neq \sigma$. If σ is of length h, let τ be the strictly increasing sequence formed by the integers not appearing in σ. Then $\rho(\xi(\tau)) \cdot \xi(\sigma')$ is ϵe (with $\epsilon = \pm 1$) if $\sigma' = \sigma$, is 0 if the length of σ' is at least equal to the length of σ and $\sigma' \neq \sigma$, and is homogeneous of degree $< r$ if the length of σ' is strictly less than that of σ. It follows that $\rho(\epsilon f \xi(\tau))$ maps $\xi(\sigma)$ upon 1 and $\xi(\sigma')$ upon 0 if $\sigma' \neq \sigma$; thus, $w = \epsilon\, \xi(\sigma_1) f\, \xi(\tau)$ has the required properties. Since the $\xi(\sigma)$ form a base of E^N, it follows immediately that $\rho(C)$ is the algebra of all vector-space endomorphisms of E^N, i.e., that $\rho(C)$ is of dimension $2^{2r} = 2^m$ equal to that of C. We conclude that ρ is a faithful representation of C, and II.2.1 is proved.

Moreover, the proof shows that the ideal Cf is a minimal left ideal of C and is identical to $E^N f$. If we observe that the elements $y_{j_1} \cdots y_{j_q} x_{i_1} \cdots x_{i_p}$ $(i_1 < \cdots < i_p \leq r, j_1 < \cdots < j_q \leq r)$ also form a base of C, we see that $fC = fE^N$ is a minimal right ideal.

We gather in the following statement the supplementary information we have obtained in the proof:

II.2.2. *The notation being as in* II.2.1, *assume further that Q is of index $r = m/2$. Let N and P be two supplementary totally singular sub-*

spaces of M, and let C^N and C^P be the subalgebras of C generated by N and P, which may be identified to the exterior algebras of these spaces. Let f be the product of the elements of a base of P. Then Cf and fC are, respectively, a minimal left ideal and a minimal right ideal; we have $Cf = C^N f$, $fC = fC^N$. Let u be in C^N: if $x \in N$, then $x(uf) = (xu)f = (x \wedge u)f$; if $y \in P$, then $y(uf) = (\delta_y \cdot u)f$, where δ_y is the antiderivation of C^N such that $\delta_y \cdot x = B(x, y)$ for $x \in N$.

Now we prove the following statement:

II.2.3. *The notation and assumptions being as in* II.2.1, *assume further that $M \neq \{0\}$; then the algebra C_+ is either simple or the direct sum of two simple ideals. The center Z of C_+ is of dimension* 2; *it is either a quadratic separable extension of K or the direct sum of two fields isomorphic with K. Assume that K is not of characteristic* 2 *and let D be the discriminant of B with respect to a base of M; then Z is spanned by* 1 *and by an element z such that $z^2 = (-1)^r D$, and z anticommutes with every element of C_-.*

Let S be a minimal left ideal of C; let ρ be the representation of C which assigns to every $u \in C$ the mapping $v \to uv$ of S into itself; then ρ is simple. Let ρ^+ be the representation of C_+ induced by ρ; among all subspaces $\neq \{0\}$ of S which are mapped into themselves by the operations of $\rho(C_+) = \rho^+(C_+)$, let S' be one of smallest possible dimension. Let x be a nonsingular element of M; then x is odd and invertible, from which it follows immediately that $C_- = xC_+ = C_+x$, $C_+ = C_-x = xC_-$. Let S'' be the transform of S' by $\rho(x)$; then it is clear from the preceding equalities that S'' is mapped into itself by all operations of $\rho^+(C_+)$ and that $S' + S''$ is mapped into itself by every operation of $\rho(C)$. Since ρ is simple, $S' + S''$ is the whole of S. If $S' \cap S'' \neq \{0\}$, then $S' \cap S'' = S'$ in virtue of the minimal character of S'; since S'' has the same dimension as S', this implies $S'' = S' = S$. If $S' \cap S'' = \{0\}$, then S is the direct sum of S', S''. Thus, ρ^+ is either simple or the sum of two simple representations. Since it is a faithful representation, C_+ is semisimple and, since any simple representation of C_+ "occurs" in any faithful representation, C_+ is simple or the sum of two simple ideals. The algebra C_+ is not central simple, because its dimension 2^{2r-1} is not a square; therefore, $Z \neq K \cdot 1$. Let K' be an algebraically closed overfield of K, let $M^{K'}$, $C^{K'}$, $C_+^{K'}$, $Z^{K'}$ be the vector space and the algebras deduced from M, C, C_+, Z by extension to K' of the basic field, and let Q' be the quadratic form on $M^{K'}$ which extends Q. Then we may regard $C^{K'}$ as the Clifford algebra of Q'; it is clear that $C_+^{K'}$ is the algebra of even elements of $C^{K'}$ and $Z^{K'}$ the center of $C_+^{K'}$. Apply the results we have just proved to $C_+^{K'}$: since $Z^{K'} \neq K' \cdot 1$, $C_+^{K'}$ is the sum of two simple ideals and $[Z^{K'}: K' \cdot 1] = 2$, since every simple algebra over K' is central simple. This proves that $[Z: K \cdot 1] = 2$. If Z is a field, then it is separable over K, since $Z^{K'}$ is semi-simple; if not, it is the direct sum of two fields isomorphic to K.

Assume now that K is not of characteristic 2. Let $(x_1, \cdots, x_m)$ be a base of M composed of mutually orthogonal vectors; set $a_i = Q(x_i)$, $z' = x_1 \cdots x_m$. We have $x_i x_j + x_j x_i = 0$ if $i \neq j$; since $m - 1$ is odd, z' anticommutes with each x_i. It follows that z' anticommutes with every element of C_- and commutes with every element of C_+. Since m is even, z' is in C_+ but obviously not in $K \cdot 1$, whence $Z = K \cdot 1 + K \cdot z'$. We easily compute z'^2 to be $(-1)^{m(m-1)/2} a_1 \cdots a_m = (-1)^r a_1 \cdots a_m$. The discriminant of B with respect to the base $(x_1, \cdots, x_m)$ is $2^{2r} a_1 \cdots a_m$; if $\mathbf{B}'$, $\mathbf{B}$ are the matrices which represent B with respect to $(x_1, \cdots, x_m)$ and to any other base $(y_1, \cdots, y_m)$ of M, then there exists an invertible matrix $\mathbf{T}$ such that ${}^t\mathbf{T} \cdot \mathbf{B}' \cdot \mathbf{T} = \mathbf{B}$ and the discriminant D of B with respect to $(y_1, \cdots, y_m)$ is $(\det T)^2 2^{2r} a_1 \cdots a_m$. Thus, there is an $a \neq 0$ in K such that $(az')^2 = (-1)^r D$, and $z = az'$ has the required properties.

Remark. The representation ρ of C which has been used in the proof of II.2.3 is simple. Since C is simple, all simple representations of C are equivalent to ρ. Thus, we see that the representation of C_+ induced by a simple representation of C is either simple or the sum of two simple representations.

II.2.4. *The notation being as in* II.2.3, *assume further that K is not of characteristic* 2 *and that C_+ is not simple. Then Z is spanned by* 1 *and by an element z_1 of square* 1 *which anticommutes with every element of C_-; the two simple ideals of C_+ are $C_+ (1 - z_1)$ and $C_+ (1 + z_1)$.*

Since $Z = K \cdot 1 + K \cdot z$ is not a field and $z^2 \in K \cdot 1$, we have $z^2 = a^2 \cdot 1$, $a \in K$; set $z_1 = a^{-1} z$. Then $z_1^2 = 1$, and $\epsilon_1 = (1 - z_1)/2$, $\epsilon_2 = (1 + z_1)/2$ are central idempotents of C_+ such that $\epsilon_1 \epsilon_2 = 0$, $\epsilon_1 + \epsilon_2 = 1$. It is clear that $\epsilon_1 \neq 0$, $\epsilon_2 \neq 0$; since C_+ is the sum of two simple ideals, these ideals are $C_+ \epsilon_1$ and $C_+ \epsilon_2$.

II.2.5. *Let the space M be represented as the direct sum of two spaces N, P each of which is in the conjugate space of the other. Let C^N, C^P be the subalgebras of C generated by N and P. Then there is a vector space isomorphism θ of the space $C^N \otimes C^P$ with C such that $\theta(u \otimes v) = uv$ for $u \in C^N$, $v \in C^P$. If K is of characteristic* 2, *then θ is also an isomorphism with respect to multiplication. Assume further that N is not isotropic and of even dimension $2r$, and let D be the discriminant of the restriction* *of B to $N \times N$ with respect to some base of N. Then C is isomorphic as an algebra to the tensor product of C^N by the Clifford algebra of the restriction of $(-1)^r DQ$ to P.*

Let $(x_1, \cdots, x_n, y_1, \cdots, y_p)$ be a base of M composed of a base $(x_1, \cdots, x_n)$ of N and a base $(y_1, \cdots, y_p)$ of P. Then the elements

$$x_{i_1} \cdots x_{i_p} = \xi(i_1, \cdots, i_p) \qquad (i_1 < \cdots < i_p \leq n)$$

form a base of C^N, the

$$y_{j_1} \cdots y_{j_q} = \eta(j_1, \cdots, j_q) \qquad (j_1 < \cdots < j_q \leq p)$$

a base of C^P, and the products $\xi(i_1, \cdots, i_p)\, \eta(j_1, \cdots, j_q)$ a base of C. It follows that there is a vector-space isomorphism θ of $C^N \otimes C^P$ with C such that $\theta(u \otimes v) = uv$ whenever u is of the form $\xi(i_1, \cdots, i_p)$ and v of the form $\eta(j_1, \cdots, j_q)$; the formula $\theta(u \otimes v) = uv$ is then true in general by linearity. Let

$$C_+^N = C^N \cap C_+ \quad C_-^N = C^N \cap C_- \quad C_+^P = C^P \cap C_+ \quad C_-^P = C^P \cap C_-.$$

We know that we may regard C^N and C^P as the Clifford algebras of the restrictions of Q to N and P; C_+^N and C_-^N are then the sets of even and odd elements of C^N, and we have similar statements for C^P. If $x \in N$, $y \in P$, then we have $B(x, y) = 0$, whence $xy + yx = 0$. It follows that every element of C_+^N commutes with every element of C^P, while an element of C_-^N anticommutes with the elements of C_-^P and commutes with those of C_+^P. If K is of characteristic 2, then every element of C^N commutes with every element of C^P, and θ is an isomorphism of algebras. Assume now that N is even-dimensional and not isotropic. Then the center of C_+^N contains an element z such that $z^2 = (-1)^r D$ which anticommutes with every element of C_-^N. It is then clear that every element of the vector space $C' = C_+^P + zC_-^P$ commutes with every element of C^N. Since z commutes with every element of C^P, we see immediately that C' is a subalgebra of C, which is generated by the space zP (for z is invertible). If $y \in P$, then we have $(zy)^2 = (-1)^r D \cdot Q(y) \cdot 1$; it follows that there exists a homomorphism φ of the Clifford algebra C of the restriction of $(-1)^r DQ$ to P onto C'. But C' and C'' are clearly both of dimension 2^p; φ is therefore an isomorphism. On the other hand, there is a homomorphism θ' of the tensor product $C^N \otimes C'$ into the algebra C such that $\theta'(u \otimes v') = uv'$ for $u \in C^N$, $v' \in C'$. The algebra $\theta'(C^N \otimes C')$ contains C^N and zP; it contains therefore N and $P = z^{-1}(zP)$; since $M = N + P$, we have $M \subset \theta'(C^N \otimes C')$ and $\theta'(C^N \otimes C') = C$. But $C^N \otimes C'$ is of dimension $2^n \cdot 2^p = 2^m = \dim C$, and θ' is therefore an isomorphism.

II.2.6. *Assume that $m = 2r + 1$ is odd and that B is nondegenerate; let D be the discriminant of B with respect to a base of N. Then the center Z of C is of dimension 2; it is spanned by 1 and by an odd element z such that $z^2 = 2(-1)^r D$. The algebra C_+ is central simple, and C is isomorphic to $Z \otimes C_+$; C is either simple or the direct sum of two simple ideals.*

Since B is nondegenerate and m odd, K is not of characteristic 2. Let x_0 be any nonsingular vector in M and N the conjugate of Kx_0; then $B(x_0, x_0) \neq 0$ and N is not isotropic. Since x_0 is invertible, $y \to x_0 y$ ($y \in N$) is a linear isomorphism of N with a subspace N' of C_+. Let C_+' be the subalgebra of C_+ generated by N'. If $y \in N$, then we have $x_0 y + y x_0 = 0$, whence $(x_0 y)^2 = -Q(x_0)\, Q(y)$. Let C'' be the Clifford algebra of the restriction of the quadratic form $-Q(x_0)\, Q$ to N; then there is a homomorphism φ of C'' onto C_+' such that $\varphi(y) = x_0 y$ for $y \in N$ (by II.1.1). But C'' is simple; φ is therefore an isomorphism. We have $\dim C_+' = \dim C'' = 2^{m-1}$. On the other hand, $u \to x_0 u$ is obviously a linear isomorphism of C_+ onto C_-; since $C = C_+ + C_-$ (direct), C_+ is of dimension 2^{m-1}. Thus, $C_+ = C_+'$ is isomorphic to C'' and is central simple. We may include x_0 in a base $(x_0, x_1, \cdots, x_{2r})$ of M composed of mutually orthogonal vectors; set $a_i = Q(x_i)$, $z_0 = x_0 x_1 \cdots x_{2r}$. We have $x_j x_i = -x_i x_j$ if $i \neq j$; since m is odd, it follows that z_0 commutes with every x_i, which proves that z is in the center of C. The discriminants of B with respect to any two bases of M differing from each other by a square factor, D is of the form $2b^2 a_0 \cdots a_{2r}$, $b \in K$. Set $z = bz_0$; then we easily see that $z^2 = 2(-1)^r D$. Moreover, 1 and z are linearly independent; $Z = K \cdot 1 + Kz$ is therefore a subalgebra of dimension 2 of the center of C, and there is a homomorphism θ of $Z \otimes C_+$ into C such that $\theta(u \otimes v) = uv$ for $u \in Z$, $v \in C_+$. Since z is odd and invertible, we obviously have $zC_+ = C_-$. Thus, $\theta(Z \otimes C_+)$ contains C_+ and C_- and is the whole of C. We have $\dim Z \otimes C_+ = 2.2^{m-1} = \dim C$, and θ is therefore an isomorphism. Since C_+ is central simple, it follows immediately that Z is the whole center of C. If $2(-1)^r D$ is not a square in K, Z is a field and C is simple: if $2(-1)^r D$ is a square in K, then, since K is not of characteristic 2, Z is the direct sum of two fields isomorphic to $K \cdot 1$ and C is the direct sum of two simple ideals.

II.2.7. *Let M' be the conjugate of M, P the set of singular vectors of M', and N a subspace of M supplementary to P. The ideal $\mathfrak{p}$ generated by P in C is in the radical of C, and $C/\mathfrak{p}$ is isomorphic to the Clifford algebra of the restriction of Q to N.*

Since M' is totally isotropic, P is a subspace of M' and $P = M'$ when K is not of characteristic 2. Let y be an element of P. Then $B(x, y) = 0$

for every $x \in M$, whence $yx = -xy$. It follows that y anticommutes with the elements of C_- and commutes with the elements of C_+. On the other hand, we have $y^2 = Q(y)\cdot 1 = 0$. If $u = u_+ + u_-$, $u_+ \in C_+$, $u_- \in C_-$, we have $uy = y(u_+ - u_-)$, whence $(yu)^2 = 0$. The elements of the left ideal yC being nilpotent, this ideal is in the radical of C, and $\mathfrak{p}$ is in the radical. Let C_N be the subalgebra of C generated by N; then $C_N + \mathfrak{p}$ is obviously a subalgebra of C containing $N + P = M$, whence $C_N + \mathfrak{p} = C$. II.2.7 will therefore be proved if we show that $C_N \cap \mathfrak{p} = \{0\}$. If K is of characteristic $\neq 2$, then $P = M'$ and the restriction of B to $N \times N$ is nondegenerate. Thus, it follows from II.2.1 and II.2.6 that C_N is semi-simple, whence $C_N \cap \mathfrak{p} = \{0\}$, since $\mathfrak{p}$ is in the radical of C. Moreover, we see that $\mathfrak{p}$ is then exactly the radical of C. Assume that K is of characteristic 2; let C_P be the algebra generated by P. Since Q is zero on P, C_P is obviously isomorphic to the exterior algebra of P and is the direct sum of $K\cdot 1$ and of the ideal $\mathfrak{p}_0$ generated by P in C_P. On the other hand, there is an isomorphism θ of $C_N \otimes C_P$ with C such that $\theta(u \otimes v) = uv$ for $u \in C_N$, $v \in C_P$ (II.2.4). Now, $C_N \otimes C_P$ is the direct sum of $C_N \otimes K\cdot 1$ and $C_N \otimes \mathfrak{p}_0$, and the latter set is the ideal generated by $\mathfrak{p}_0$ in C. It follows immediately that $\theta(C_N \otimes \mathfrak{p}_0) = \mathfrak{p}$ and that C is the direct sum of C_N and $\mathfrak{p}$.

The notation being as in II.2.7, let R be a subspace of M' supplementary to P in M'. We may assume that $R \subset N$; let S be a subspace of N supplementary to R. Then M is the direct sum of S and M'. Assume that K is of characteristic 2 (otherwise $R = \{0\}$). The restriction of B to $S \times S$ being nondegenerate, S is even-dimensional and the algebra C_S generated by S is central simple. If C_R is the algebra generated by R, C_N is isomorphic to $C_S \otimes C_R$. Let us now consider the structure of C_R. Let $\{x_1, \cdots, x_d\}$ be a base of R, and $Q(x_i) = a_i$ $(1 \leq i \leq r)$. We have $x_ix_j + x_jx_i = B(x_i, x_j)\cdot 1 = 0$, and C_R is a commutative algebra. Let L be the subfield of an algebraic closure of K obtained by adjunction of $a_1^{1/2}, \cdots, a_d^{1/2}$ to K. We may assume that $L = K(a_1^{1/2}, \cdots, a_e^{1/2})$, e being an integer $\leq d$, and that $[L:K] = 2^e$. Each a_i $(1 \leq i \leq d)$ is the square of an element $a_i^{1/2}$ of L; if $u_1, \cdots, u_d$ are in K, we have, in L,

$$\left(\sum_{i=1}^{d} u_i a_i^{1/2}\right)^2 = \sum_{i=1}^{d} u_i^2 a_i = Q\left(\sum_{i=1}^{d} u_i x_i\right).$$

Therefore, it follows from II.1.1 that there exists a homomorphism ψ of C_R onto L such that $\psi(x_i) = a_i^{1/2}$ $(1 \leq i \leq d)$. Let $R' = Kx_1 + \cdots + Kx_e$, and let $C_{R'}$ be the algebra generated by R'; thence $C_{R'}$ is of dimension $2^e = [L:K]$, and $\psi(C_{R'})$, which contains $a_i^{1/2}$ for $1 \leq i \leq e$, is the whole of L. This shows that ψ is an isomorphism of $C_{R'}$ with L. Let us assume from now on that $C_{R'} = L$. If $i > e$, then there is a $z_i \in L$ such that $z_i^2 = x_i^2 = a_i \cdot 1$; it follows that $(x_i - z_i)^2 = 0$. The elements $x_i - z_i$ $(i > e)$ generate a nilpotent ideal $\mathfrak{N}$ of C_R, and it is clear that every element of C_R is congruent modulo $\mathfrak{N}$ to some element in L. This shows that the kernel of ψ is $\mathfrak{N}$ and that $\mathfrak{N}$ is the radical. Thus, we see that the quotient of C_R by its radical is a field, purely inseparable of exponent 1 over K.

2.3. The Group of Clifford

We shall now assume that the bilinear form B is nondegenerate, i.e., that Q is of rank m and defect 0. In particular, if K is of characteristic 2, m is even. We denote by C the Clifford algebra of Q and by G the orthogonal group of Q.

We shall call *Clifford group* of G, and denote by Γ, the group of invertible elements s of C such that $sxs^{-1} \in M$ for every $x \in M$. If $s \in \Gamma$, we shall denote by $\chi(s)$ the linear automorphism $x \to sxs^{-1}$ of M. It is clear that χ is a linear representation of Γ; we shall call it the *vector representation* of Γ, to distinguish it from the spin representation to be introduced later.

Let s be in Γ. Then we have, for $x \in M$, $Q(sxs^{-1}) \cdot 1 = (sxs^{-1})^2 = sx^2s^{-1} = Q(x) \cdot 1$. It follows that χ maps Γ into the orthogonal group G of Q.

II.3.1. *If m is even, then $\chi(\Gamma) = G$. If m is odd, then $\chi(\Gamma)$ is the group G^+ of operations of determinant* 1 *in G. If x is any nonsingular element of M, then $x \in \Gamma$ and $\chi(x)$ is the mapping $y \to -\tau \cdot y$, where τ is the symmetry with respect to the conjugate hyperplane of Kx. Let Z^* be the multiplicative group of invertible elements of the center Z of C; then Z^* is the kernel of χ and, except in the case where K is a field with* 2 *elements,* dim $M = 4$ *and Q is of index* 2, *$Z^* \cup (\Gamma \cap M)$ is s set of generators of the group Γ.*

Let σ be any operation in G. Then we have $(\sigma \cdot x)^2 = Q(\sigma x) \cdot 1 = Q(x) \cdot 1$ for $x \in M$, and it follows from II.1.1 that σ may be extended to an automorphism σ' of the algebra C. If σ' leaves the elements of the center Z of C fixed, then σ' is an inner automorphism. This follows from the Noether-Skolem theorem if C is simple. If not, then C is the direct sum of two simple ideals $\mathfrak{a}_1$ and $\mathfrak{a}_2$; $\mathfrak{a}_i$ is generated by a central idempotent e_i and is central simple. It follows that σ' transforms $\mathfrak{a}_i$ into itself and that its restriction to $\mathfrak{a}_i$ is an inner automorphism produced by an element s_i of $\mathfrak{a}_i$ which is invertible in $\mathfrak{a}_i$; since $s_1 s_2 = 0$, $s = s_1 + s_2$ is invertible in C and σ' is the inner automorphism produced by s.

Now, if m is even, C is central simple, and σ' is an inner automorphism $u \to sus^{-1}$, s being invertible in C. Since $\sigma'(M) = M$, s belongs to Γ and $\chi(s) = \sigma$.

Let ζ be the mapping $x \to -x$ ($x \in M$), and ζ' the automorphism of C which extends ζ. It is clear that $\zeta'(u) = -u$ for any $u \in C_-$. Now, if m is odd, then Z contains an odd element $z \neq 0$ (II.2.6) and K is not of characteristic 2. Since $\zeta'(z) = -z$, ζ' does not leave the elements of Z fixed. Were ζ in $\chi(\Gamma)$, there would exist an $s \in \Gamma$ such that $\zeta' \cdot x = sxs^{-1}$ for all $x \in M$. Since M generates C, we would have $\zeta' \cdot u = sus^{-1}$ for all $u \in C$, which is not the case. Thus, if m is odd, ζ does not belong to $\chi(\Gamma)$ and $\chi(\Gamma) \neq G$.

Let x be a nonsingular element of M. Then x is invertible, and $x^{-1} = (Q(x))^{-1}x$. We have $xy + yx = B(x, y) \cdot 1$ for $y \in M$, whence

$$xyx^{-1} = (Q(x))^{-1} B(x, y)x - y = -\tau \cdot y,$$

where τ is the symmetry with respect to the conjugate of $K \cdot x$. It follows that $x \in \Gamma$ and that $\chi(x) = -\tau$. It is clear that $Z^* \subset \Gamma$, and that Z^* is in the kernel of χ. Conversely, if $s \in \Gamma$, $\chi(s) = 1$, then s commutes with every element of M, and $s \in Z \cap \Gamma = Z^*$. Assume that we are not considering the exceptional case mentioned in the statement. Any operation σ of G may be written as a product $\tau_1 \cdots \tau_h$ of symmetries with respect to hyperplanes whose conjugates contain nonsingular vectors $x_1, \cdots, x_h$ (by I.5.1). Thus, since $\zeta^2 = 1$, we have $\sigma = \zeta^h \chi(x_1 \cdots x_h)$. If m is odd, then we have det $\tau_i = +1$, det $\zeta = -1$, and, if $\sigma \in G^+$, then we have $h \equiv 0 \pmod 2$ and $\sigma = \chi(x_1 \cdots x_h)$; if $\sigma = \chi(s)$, $s \in \Gamma$, then $s = s_0 x_1 \cdots x_h$ with $s_0 \in Z^*$. If m is even, then ζ belongs to the group generated by $\Gamma \cap M$. This is obvious if K is of characteristic 2, ζ being then the identity; if not, let $(y_1, \cdots, y_m)$ be a base of M composed of mutually orthogonal vectors. Then $\chi(y_i) \cdot y_j$ is $-y_j$ if $i \neq j$, and $\chi(y_i) \cdot y_i = y_i$, whence $\zeta = \chi(y_1 \cdots y_m)$. Thus, it follows in the same way as above that any $s \in \Gamma$ belongs to the group generated by $\Gamma \cap M$ and Z^*.

II.3.2. *Every $s \in \Gamma$ may be written in the form zs', where z is in the center of C and s' is an element of Γ which is either even or odd. If m is even, s is either even or odd.*

If we are not in the exceptional case of II.3.1, then s is the product of an element of Z^* by a certain number of elements of $\Gamma \cap M$, which proves our assertion in that case. If m is even, we may also use the following argument, which applies even in the exceptional case. If $\sigma = \chi(s)$, then we have $sx = (\sigma\cdot x)s$ for all $x \in M$. Let $s = s_+ + s_-$, $s_+ \in C_+$, $s_- \in C_-$. Since x and $\sigma\cdot x$ are odd, we have $s_+x = (\sigma\cdot x)s_+$, and $s^{-1}s_+$ commutes with every $x \in M$ and belongs therefore to the center $K\cdot 1$ of C. If $s_+ = as$, $a \in K$, and if $a \neq 0$ then s is even; if $a = 0$, then $s = s_-$ is odd.

We shall denote by Γ^+ *the group of even elements of* Γ.

II.3.3. *If* $m > 0$, *the group* $\chi(\Gamma^+)$ *is a subgroup of index* 2 *of* G. *If* K *is not of characteristic* 2, *then* $\chi(\Gamma^+)$ *is the group of operations of determinant* 1 *in* G.

Since $m > 0$, M contains a nonsingular vector x; x is an odd element of Γ, whence $\Gamma \neq \Gamma^+$. If m is even, then the center of C is $K\cdot 1$, which is in C_+; it follows immediately that $\chi(\Gamma^+)$ is then of index 2 in G. If m is odd, then the center of C contains an invertible odd element (by II.2.6); this element is in Γ but not in Γ_+. Thus, it follows from II.3.2 that every element of Γ is the product of an element of the center of C by an element of Γ^+, whence $\chi(\Gamma) = \chi(\Gamma^+)$. This group is the group of operations of determinant 1 in G. The determinant of any element of G is ± 1, and G contains an operation of determinant -1, for instance, the mapping $x \to -x$. It follows that $\chi(\Gamma^+)$ is of index 2 in G. Now, assume that m is even and that the characteristic of K is $\neq 2$. Any $s \in \Gamma$ is representable in the form $cx_1 \cdots x_h$, $c \in K$, $x_i \in \Gamma \cap M$ $(1 \leq i \leq h)$, and it is easily seen that $\det \chi(x_i) = -1$ $(1 \leq i \leq h)$, whence $\det \chi(s) = (-1)^h$. Thus, $\det \chi(s)$ is 1 or -1 according as to whether s is in Γ^+ or not, and this completes the proof of II.3.3.

The group Γ^+ will be called the *special Clifford group of* Q; the group $\chi(\Gamma^+)$ will be called the *special orthogonal group of* Q or also the *group of rotations* (its elements being called rotations). If K is of characteristic 2, then every operation of G is of determinant 1, but G^+ is then still of index 2 in G.

II.3.4. *If* m *is even* > 0, *the group* Γ^+ *is generated by the products of two nonsingular elements of* M *except in the case where* K *has* 2 *elements,* $m = 4$, *and* Q *is of index* 2. *If* m *is odd, let* z *be an odd invertible element of the center of* C. *Then* Γ^+ *is generated by the products* xz, *where* x *runs over the nonsingular elements of* M.

The assertion relative to the case m even follows immediately from II.3.1. Assume m odd; it is clear that, for any nonsingular $x \in M$, xz belongs to Γ^+. Let, conversely, s be in Γ^+; then, by II.3.1, s may be written as $s = \zeta x_1 \cdots x_h$, where ζ is in the center of C and $x_i \in M$.

The center of C is $K\cdot 1 + K\cdot z$ (II.2.6); if h is even, then $\zeta \in K\cdot 1$, while, if h is odd, $\zeta \in K\cdot z$. The element z^2, which is an even element of the center of C, is of the form $a\cdot 1$, $a \in K$, $a \neq 0$. If $h = 2h'$, then we may write

$$s = (\zeta a^{-h'} x_1 z)(x_2 z) \cdots (x_h z);$$

if $h = 2h' + 1$, let $\zeta = cz$. Then we have

$$s = a^{-h'}(cx_1 z) \cdots (x_{h-1} z)(x_h z);$$

this concludes the proof of II.3.4.

It is easily seen that the case where K has 2 elements, $m = 4$, and Q is of index 2 is actually an exceptional case.

We shall now take into consideration the main antiautomorphism α of the algebra C which has been introduced in II.1. We know that $\alpha(x) = x$ for $x \in M$; it follows that $\alpha(C_+) = C_+$, $\alpha(C_-) = C_-$.

II.3.5. *If s is any element of the Clifford group Γ, then $\alpha(s) \in \Gamma$ and $\alpha(s)s$ is an element of the center of C; if $s \in \Gamma^+$, then $\alpha(s)s \in K\cdot 1$.*

Let $\sigma = \chi(s)$; then, for $x \in M$, we have $sx = (\sigma\cdot x)s$, whence $x\alpha(s) = \alpha(s)\sigma\cdot x$ and $\alpha(s)sx = x\alpha(s)s$, which shows that $\alpha(s)s$ is in the center of C. This element being obviously invertible, it follows that $\alpha(s) \in \Gamma$. If s is even, then so is $\alpha(s)s$, and $\alpha(s)s \in K\cdot 1$.

Let s and t be in Γ. Then we have $\alpha(st)st = \alpha(t)\alpha(s)st = \alpha(s)s\alpha(t)t$, which shows that $s \to \alpha(s)s$ is a homomorphism of Γ into the multiplicative group of invertible elements of the center of C. Whenever $\alpha(s)s$ is in $K\cdot 1$, we shall set $\alpha(s)s = \lambda(s)\cdot 1$, $\lambda(s) \in K$. This always happens if $s \in \Gamma^+$. It also happens if $s \in \Gamma \cap M$, for then $\lambda(s) = Q(s)$; thus, $\lambda(s)$ is always defined for all $s \in \Gamma$ if m is even. We have

$$\lambda(c\cdot 1) = c^2, \quad \text{if} \quad c \in K, \quad c \neq 0;$$

$$\lambda(x) = Q(x), \quad \text{if} \quad x \in M,\ Q(x) \neq 0.$$

The element $\lambda(s)$ (when it is defined) will be called the *norm* of s, and λ will be called the *norm homomorphism.*

We shall denote by Γ_0 the group of elements $s \in \Gamma$ such that $\alpha(s)s = 1$, and by Γ_0^+ the group $\Gamma_0 \cap \Gamma^+$. We shall call Γ_0^+ the reduced Clifford group of Γ; the group $\chi(\Gamma_0^+)$ will be called the reduced orthogonal group of Q and will be denoted by G_0^+.

The group Γ/Γ_0 is clearly abelian, which shows that the commutator subgroup Γ' of Γ is contained in Γ_0. If s is an element of Γ such that $\chi(s)$ is not in G^+, then $\chi(\Gamma)$ is generated by $\chi(\Gamma^+)$ and $\chi(s)$, and Γ is generated by Γ^+, s, and its center. It follows that Γ/Γ^+ is abelian and that

$\Gamma' \subset \Gamma^+$, whence $\Gamma' \subset \Gamma_0^+$. If m is even, then $G = \chi(\Gamma)$ and we see that the commutator subgroup G' of G is in G_0^+. If m is odd, then the center of G contains an element ζ not in G^+ (namely, the mapping $x \to -x$), and $G = G^+ \cup (\zeta G^+)$, which shows that G' is also the commutator subgroup of G^+; since $G^+ = \chi(\Gamma^+)$, we see that, here again, $G' \subset G_0^+$.

III.3.6. *Let H be the subgroup of the multiplicative group $K_\bullet$ of elements $\neq 0$ in K which is generated by the products $Q(x)Q(y)$, x and y running over the nonsingular elements of M, and let $K_\bullet^2$ be the group of squares of elements of $K_\bullet$. If $m > 0$, the group G^+/G_0^+ is isomorphic to $H/K_\bullet^2$.*

The kernel of the restriction of χ to Γ^+ is $K_\bullet \cdot 1$. Since any element of this kernel is an even element of the center of C, G^+/G_0^+ is isomorphic to $\Gamma^+/K_\bullet\Gamma_0^+$, i.e., also to $\lambda(\Gamma^+)/\lambda(K_\bullet)$. If K has only 2 elements, then $H = K_\bullet = K_\bullet^2 = \lambda(\Gamma^+)$ and II.3.6 is obvious. Assume that this is not the case. We have $\lambda(K_\bullet) = K_\bullet^2$; if m is even, then Γ^+ is generated by the products xy, for $x, y \in \Gamma \cap M$, whence $\lambda(\Gamma^+) = H$. Assume m odd, and let $(x_1, \cdots, x_m)$ be a base of M composed of mutually orthogonal vectors. Then $z = x_1 \cdots x_m$ is an odd invertible element of the center of C, and Γ^+ is generated by the products xz, with $x \in \Gamma \cap M$ (II.3.4). We have $\lambda(xz) = \lambda(zx) = (Q(x_1)\,Q(x_2)) \cdots (Q(x_{m-2})\,Q(x_{m-1}))\,(Q(x_m)\,Q(x))$ and $\lambda(\Gamma^+) = H$; II.3.6 is thereby proved.

II.3.7. *If the index of Q is > 0, then G^+/G_0^+ is isomorphic to $K_\bullet/K_\bullet^2$.*

This follows from II.3.6, since Q then assumes all values in K (by I.3.3).

II.3.8. *Assume that the index of Q is > 0 and that we are not in the following exceptional case: K has 2 elements, $\dim M = 4$, and Q is of index 2. Then G_0^+ is the group of commutators of G.*

Since Q is of index > 0, there exist two singular vectors x, y such that $B(x, y) = 1$. The plane $P = Kx + Ky$ is not isotropic; let P' be its conjugate. We shall prove that every $\sigma \in G$ is the product of an element of the commutator subgroup G' of G and of an operation which leaves the elements of P' fixed. We first consider the case where σ is the symmetry with respect to a hyperplane H whose conjugate contains a nonsingular vector z. Let $z_1 = x + Q(z)y$, whence $Q(z) = Q(z_1)$. There is a $\tau \in G$ such that $\tau \cdot z = z_1$ (I.4.1); we may write $\sigma = (\sigma\,\tau\,\sigma^{-1}\,\tau^{-1})(\tau\,\sigma\,\tau^{-1})$, and $\tau\sigma\tau^{-1}$ is the symmetry with respect to the conjugate H_1 of Kz_1. Since $P' \subset H_1$, $\tau\sigma\tau^{-1}$ leaves the elements of P' fixed, which proves our assertion in that case. To establish it in the general case, it will be sufficient (in virtue of I.5.1) to show that, if our assertion is true of σ, σ', then it is also true of $\sigma\sigma'$. We have $\sigma = \sigma_1\sigma_2$, $\sigma' = \sigma'_1\sigma'_2$, where σ_1, $\sigma'_1 \in G'$ and σ_2, σ'_2 leave the elements of P' fixed; thus, we have $\sigma\sigma' = (\sigma_1\sigma_2\sigma'_1\sigma_2^{-1})(\sigma_2\sigma'_2)$, which proves our assertion for $\sigma\sigma'$. Now, let $\sigma = \sigma_1\sigma_2$ be in G_0^+, $\sigma_1 \in G'$, σ_2, leaving the elements of P' fixed. Since $G' \subset G_0^+$, we have $\sigma_2 \in G_0^+$. The only singular vectors of P are those of Kx and Ky; thus, $\sigma_2 \cdot x$ is either in Kx or in Ky. Moreover, it is clear that any operation of G_0^+ which leaves x and the elements of P' fixed is the identity. We shall see that it is impossible that $\sigma \cdot x = ay$, $a \in K$. For, let then τ be the symmetry with respect to the conjugate hyperplane of $K(x + ay)$. (We have $a \neq 0$, whence $Q(x + ay) \neq 0$.) Then it is easily seen that $\tau \cdot x = ay$, and τ leaves the element of P' fixed. Since τ is not in G^+, $\sigma_2 \neq \tau$, which proves our assertion. Therefore, we have $\sigma_2 \cdot x = ax$, $a \in K$, whence $\sigma_2 \cdot y = a^{-1}y$. Let $s = a \cdot 1 + (1 - a)yx$; any element of P' anticommutes with every element of P and commutes therefore with s. We have

$$(a \cdot 1 + (1 - a)yx)\,(a \cdot 1 + (1 - a)xy) = a \cdot 1,$$

so that

$$s^{-1} = a^{-1}\,(a \cdot 1 + (1 - a)xy).$$

We have $sxs^{-1} = ax$ and, since $s = 1 + (1 - a)xy$, $sys^{-1} = a^{-1}y$. This shows that $s \in \Gamma^+$ and $\chi(s) = \sigma_2$. On the other hand, there is an $s' \in \Gamma_0^+$ such that $\chi(s') = \sigma_2$; it follows that $s's^{-1}$ is in the center of Γ^+, i.e., that $s = cs'$, c a scalar. We have $\alpha(s) = a \cdot 1 + (1 - a)xy$, whence $c^2 = \lambda(s) = a$. Now let σ_3 be the operation of G which maps x upon cx, y upon $c^{-1}y$, and the elements of P' upon themselves, and let τ be the operation of G which exchanges x and y and maps the elements of P' upon themselves. Then we see that $\tau\sigma_3^{-1}\tau^{-1}\sigma_3 \cdot x = c^2x = \sigma_2 \cdot x$, whence $\sigma_2 = \tau\sigma_3^{-1}\tau^{-1}\sigma_3 \in G'$. Thus, we have proved that $G_0^+ \subset G'$. Since $G' \subset G_0^+$, II.3.8 is proved.

II.3.9. *The assumptions being as in* II.3.8, *assume furthermore that* $\dim M > 2$. *Then* G_0^+ *is also the commutator subgroup of* G^+.

We have only to prove that the commutator subgroup G' of G is contained in the commutator subgroup H of G^+. It is clear that H is a normal subgroup of G. Let σ_1, σ_2, τ be in G; the formula

$$(\sigma_1\sigma_2)\tau(\sigma_1\sigma_2)^{-1}\tau^{-1} = \sigma_1(\sigma_2\tau\sigma_2^{-1}\tau^{-1})\sigma_1^{-1}(\sigma_1\tau\sigma_1^{-1}\tau^{-1})$$

shows that, if the commutators of σ_1, τ and of σ_2, τ are in H, then so is the commutator of $\sigma_1\sigma_2$, τ. Every element of G^+ is a product of symmetries with respect to hyperplanes (whose conjugates contain nonsingular vectors). It will therefore be sufficient to prove that, if σ is such a symmetry and $\tau \in G$, then $\sigma\tau\sigma^{-1}\tau^{-1} \in H$, or, which amounts to the same, that $\tau\sigma\tau^{-1}\sigma^{-1} \in H$. Decomposing τ into symmetries, we are reduced to consider the case where σ and τ are symmetries with respect to hyperplanes whose conjugates contain nonsingular vectors x and y. The conjugate of Ky is obviously not totally singular; let y' be a nonsingular vector of this conjugate and σ' the symmetry with respect to the conjugate hyperplane of Ky'. Let $\zeta = \sigma\sigma'$; then $\sigma = \zeta\sigma'$ and, since σ' commutes with τ, it follows from the formula written above that it is sufficient to show that $\zeta\tau\zeta^{-1}\tau^{-1} \in H$. Assume first that the conjugate space P of $Kx + Ky'$ is not totally singular, and then let x' be a nonsingular vector of P and let τ' be the symmetry with respect to the conjugate of Kx'. Then τ' commutes with σ and σ', and therefore with ζ. We write $\tau = \tau\tau'\cdot\tau'$; since τ' commutes with ζ and $\tau\tau' \in G^+$, the commutators of ζ and τ and of ζ and $\tau\tau'$ are in H, which shows that the commutator of ζ and τ is in H. Assume now that P is totally singular. Then we have $P \subset Kx + Ky'$ and, since Kx is not singular, $P \neq Kx + Ky'$. If $m = \dim M$, then P is of dimension $m - 2$. Since $m > 2$, P is of dimension 1 and $m = 3$. But, if m is odd, then the center of G contains an element not in G^+ (namely, the mapping $x \rightarrow -x$), from which it follows immediately that G and G^+ have the same commutator subgroup.

2.4. Spinors (Even Dimension)

We assume that the space M is of even dimension $m = 2r$, and that B is nondegenerate. We denote by G the orthogonal group of Q, by C its Clifford algebra, by C_+, C_- the spaces of even and odd elements of C, by Γ the Clifford group of Q, by Γ^+ its special Clifford group, and by Γ_0^+ its reduced Clifford group.

We know that all simple representations of the simple algebra C are equivalent. We select one of them, say ρ, and we call the space S of this representation the *space of spinors* of Q. The representation ρ of C is called the *spin representation* of C; the representation ρ^+ of C_+ induced by ρ is called the spin representation of C_+. The representation ρ of C induces a representation of Γ, which will still be denoted by ρ; it also induces representations of Γ^+ and Γ_0^+ which are denoted by ρ^+, ρ_0^+; all these representations are also called *spin representations.*

II.4.1. *Except in the case where K has 2 elements, $m = 2$, and Q is of index 1, Γ is a set of generators of the algebra C and the spin representation of Γ is simple.*

We first establish the following:

Lemma 1. *Let R be a finite-dimensional vector space over a field K and Q_1 a quadratic form on R whose associated bilinear form B_1 is nondegenerate. Let x_1 be any element $\neq 0$ of R and N the subspace of R spanned by all vectors x such that $Q(x) = Q(x_1)$. Then we have $N = R$ unless R is of dimension 2, Q is of index 1, and K has either 2 or 3 elements.*

It is obvious that N is mapped into itself by the operations of the orthogonal group G_1 of the form Q_1. Lemma 1 therefore follows from I.6.2 and I.6.7.

Now, Γ contains every nonsingular vector of M. If $m = 2$ and K has 3 elements, we see immediately that there exist two linearly independent nonsingular vectors in M. Thus, if we are not considering the exceptional case of II.4.1, then M is spanned by $\Gamma \cap M$, which shows that Γ generates C. Since the spin representation of C is simple, so is the spin representation of Γ. If we are considering the exceptional case, then M is spanned by two singular vectors x and y such that $B(x, y) = 1$. Then $\Gamma = \{1, x + y\}$, and it is easily seen that the spin representation of Γ is not simple.

Consider now the representation ρ^+ of C_+. This representation is either simple or the sum of two simple representations (see the remark which follows the proof of II.2.3). If C_+ is not simple, then C_+ has two *inequivalent* simple representations, and both must occur in ρ^+, since ρ^+ is faithful. In that case, ρ^+ is the sum of two inequivalent simple representations. It follows that S may be represented in one and only one way as the sum of two subspaces each of which yields a simple representation of C_+. These two spaces are then called the spaces of *half-spinors*, and the corresponding representations of C_+ the *half-spin representations*. The representations of Γ^+, Γ_0^+ induced by the half-spin representations of C_+ are called the half-spin representations of these groups.

II.4.2. *The spin representation ρ^+ of Γ^+ is either simple or the sum of two simple representations. If C_+ is not simple and if we are not in the exceptional case of* II.4.1, *then the half-spin representations of Γ^+ are simple and inequivalent to each other.*

We have seen in the proof of II.4.1 that, if we are not considering the exceptional case, M is spanned by its nonsingular vectors. On the other hand, C_+ is generated by all products of 2 elements of M and therefore also (outside the exceptional case) by the products of two nonsingular vectors of M. But these products are in Γ^+, and Γ^+ is therefore a set of generators of the algebra C_+. In the exceptional case of II.4.1, we have $\Gamma^+ = \{1\}$ and the spin representation splits into two simple representations; II.4.2 is thereby proved.

II.4.3. *The spin representation ρ_0^+ of Γ_0^+ is either simple or the sum of two simple representations; if the spin representation of Γ^+ is simple, then so is ρ_0^+. If C_+ is not a simple algebra, then the half-spin representations of Γ_0^+ are simple; they are inequivalent to each other except if $m = 2$, Q is of index 1, and K has either 2 or 3 elements.*

We may assume $M \neq \{0\}$; let x_1 be a nonsingular vector in M and $a_1 = Q(x_1)$. Assume that M is spanned by the set of all vectors x such that $Q(x) = Q(x_1)$. Since C_+ is generated by all products of two elements of M, it is also generated by the elements of the form $a_1^{-1}xy$, where x, y are vectors such that $Q(x) = Q(y) = a_1$. But $a_1^{-1}xy$ then belongs to Γ_0^+, since $\lambda(a_1^{-1}xy) = a_1^{-2}\,Q(x)\,Q(y)$; thus, Γ_0^+ is in that case a set of generators of C_+. If the set of vectors x such that $Q(x) = a$ does not span M, then $m = 2$, Q is of index 1, and K has either 2 or 3 elements. In these cases, it is easily seen that the representation ρ^+ of Γ^+ is never simple: it splits into two representations of degree 1. Since every representation of degree 1 is simple, II.4.3 is proved.

2.5. Spinors (Odd Dimension)

We assume now that the space M is of odd dimension $m = 2r + 1$ and that B is nondegenerate. Otherwise, we use the same notation as in Section 4.

The algebra C_+ is now central simple (II.2), and its simple representations are all equivalent to each other. We select one, say ρ^+, which we call the *spin representation*; the space S of this representation will be called the *space of spinors*. The representations of Γ^+, Γ_0^+ induced by ρ^+ are called the *spin representations* of these groups.

II.5.1. *The group Γ_0^+ is a set of generators of the algebra C_+; the spin representations of Γ^+, Γ_0^+ are simple.*

Let x_1 be a nonsingular vector in M. Then M is spanned by the vectors $Q(x)$ such that $Q(x) = Q(x_1)$ (Lemma 1, II.4). It follows as in the proof of II.4.3 that Γ_0^+ is a set of generators of the algebra C_+. The second assertion of II.5.1 follows immediately from the first.

II.5.2. *If the algebra C is not simple, then it is possible in exactly two ways to extend the spin representation of C_+ to a representation of the algebra C.*

The center Z of C is spanned by 1 and by an odd element z such that $z^2 \in K \cdot 1$. Since C is not simple, z^2 must be a square in K, and we may assume without loss of generality that $z^2 = 1$. Any $u \in C$ is uniquely representable in the form $u = u_1 + u_2 z$, where u_1, u_2 are in C_+. Since z is in the center of C, the mappings $\varphi: u \rightarrow u_1 + u_2$ and $\varphi': u \rightarrow u_1 - u_2$ are homomorphisms of C into C_+; the representations $\rho = \rho^+ \circ \varphi$, $\rho' = \rho^+ \circ \varphi'$ are representations of C which extend ρ^+. Conversely, let ρ'' be any representation of C which extends ρ^+. Let $\sigma = \rho''(z)$; then σ^2 is the identity mapping I of the space S of spinors, and σ commutes with every operation of $\rho^+(C_+)$. The space S is the sum of the space S_1 of elements w such that $\sigma \cdot w = w$ and of the space S_2 of elements w' such that $\sigma \cdot w' = -w'$. But these spaces are mapped into themselves by the operations of $\rho^+(C_+)$. Since ρ_+ is simple, one of S_1, S_2 is S and the other $\{0\}$, whence $\sigma = \pm I$. It follows that ρ'' is one of the representation ρ, ρ'.

If C is not simple, then the two representations of C which extend ρ^+ are called the two *spin representations of* C; the representations of Γ induced by these spin representations are called the *spin representations of* Γ.

2.6. Imbedded Spaces

We shall assume that B is nondegenerate. We shall denote by $\overline{M}$ a non-isotropic subspace of M, by $\overline{Q}$ the restriction of Q to $\overline{M}$, by C, $\overline{C}$ the Clifford algebras of Q, $\overline{Q}$, by C_+, $\overline{C}_+$ the algebras of even elements of C, $\overline{C}$, by Γ, $\overline{\Gamma}$ the Clifford groups of Q, $\overline{Q}$, by Γ^+, $\overline{\Gamma}^+$ their special Clifford groups, by $\Gamma_0{}^+$, $\overline{\Gamma}_0{}^+$ their reduced Clifford groups.

We shall identify $\overline{C}$ to the subalgebra of C generated by $\overline{M}$; we then have $\overline{C}_+ = \overline{C} \cap C_+$.

II.6.1. *The group $\overline{\Gamma}^+$ is a subgroup of Γ^+; if $\bar{s} \in \overline{\Gamma}^+$, then the norm of $\bar{s}$ is the same whether we consider $\bar{s}$ as an element of Γ^+ or of $\overline{\Gamma}^+$, and $\overline{\Gamma}_0{}^+ = \overline{\Gamma}^+ \cap \Gamma_0{}^+$. If $\overline{M}$ is of even dimension, then $\overline{\Gamma} \subset \Gamma$ and any element of $\overline{\Gamma}$ has the same norm in Γ as in $\overline{\Gamma}$.*

Let N be the conjugate space of $\overline{M}$. If $y \in N$, then y anticommutes with every element of $\overline{M}$; it follows that y anticommutes with every odd element of $\overline{C}$ and commutes with every element of $\overline{C}_+$. If $\bar{s} \in \overline{\Gamma}$ is either even or odd, then we have $\bar{s}y\bar{s}^{-1} = \pm y$ and $\bar{s}N\bar{s}^{-1} = N$. Since $M = \overline{M} + N$, $\bar{s}M\bar{s}^{-1} = M$, $\bar{s}$ is in Γ. This shows that $\overline{\Gamma}^+ \subset \Gamma^+$ and that $\overline{\Gamma} \subset \Gamma$ if $\overline{M}$ is of even dimension (see II.3.2). It is obvious that the main antiautomorphism of C induces the main antiautomorphism of $\overline{C}$; the remaining statements of II.6.1 follow immediately from this observation.

II.6.2. *Let χ and $\bar{\chi}$ be the vector representations of Γ and $\bar{\Gamma}$; if $\bar{s} \in \Gamma \cap \bar{\Gamma}$, then $\bar{\chi}(\bar{s})$ is the restriction of $\chi(\bar{s})$ to $\bar{M}$, and, if $\bar{s} \in \bar{\Gamma}^+$, then $\chi(\bar{s})$ leaves fixed the elements of the conjugate space of $\bar{M}$. The representation of $\bar{\Gamma}^+$ induced by the spin representation of Γ^+ is the sum of a certain number of representations equivalent to the spin representation of $\bar{\Gamma}^+$. Assume now that C_+ is not simple and that $\bar{M} \neq M$. Then the representation of $\bar{\Gamma}^+$ induced by a half-spin representation of Γ^+ is the sum of a certain number of representations equivalent to the spin representation of $\bar{\Gamma}^+$.*

If $\bar{s} \in \Gamma \cap \bar{\Gamma}$, then we have $\chi(\bar{s})\cdot x = \bar{s}x\bar{s}^{-1}$ for all $x \in \bar{M}$, which shows that the restriction of $\chi(\bar{s})$ to $\bar{M}$ is $\bar{\chi}(\bar{s})$. Every element y of the conjugate space of $\bar{M}$ anticommutes with every element of $\bar{M}$ and therefore commutes with every element of $\bar{C}_+$, which shows that $\chi(\bar{s})\cdot y = y$ if $\bar{s} \in \bar{\Gamma}_+$. We have $\bar{C}_+ \subset C_+$; if $\bar{C}_+$ is simple, then the representation of $\bar{C}_+$ induced by a representation of C_+ is the sum of a certain number of representations all equivalent to the spin representation of $\bar{C}_+$. (Observe that the unit element of $\bar{C}_+$ is also unit element of $C_+\cdot$) This shows that the representation of $\bar{\Gamma}^+$ induced by the spin representation of Γ^+ (or by a half-spin representation of Γ^+, if C_+ is not simple) is the sum of a certain number of representations equivalent to the spin representation of $\bar{\Gamma}^+$. Assume now that $\bar{C}_+$ is not simple but that C_+ is, and let τ be the representation of $\bar{C}_+$ induced by the spin representation of C_+. Let ω_1 and ω_2 be the two half-spin representations of $\bar{C}_+$; then the spin representation of $\bar{C}_+$ is $\omega_1 + \omega_2$. The representation τ is the sum of a certain number of simple representation of $\bar{C}_+$, each one of which is equivalent to ω_1 or ω_2; we wish to prove that ω_1 and ω_2 occur the same number of times in τ. We may obviously assume $\bar{M} \neq M$. The regular representation of C_+ on itself is the sum of a certain number of representations all equivalent to the spin representation; it will therefore be sufficient to prove that ω_1 and ω_2 occur the same number of times in the representation θ of $\bar{C}_+$ induced by the regular representation of C_+.

The algebra $\bar{C}_+$ is the sum of two simple ideals $\bar{\mathfrak{a}}_1$ and $\bar{\mathfrak{a}}_2$. Let $\bar{x}$ be a nonsingular element of $\bar{M}$; then $\bar{u} \to \bar{x}\bar{u}\bar{x}^{-1}$ is an automorphism j of $\bar{C}_+$. We assert that j exchanges the ideals $\bar{\mathfrak{a}}_1$ and $\bar{\mathfrak{a}}_2$. We may write $\bar{\mathfrak{a}}_i = \bar{C}_+\epsilon_i$, where ϵ_i is a central idempotent of $\bar{C}_+$. If we had $j(\mathfrak{a}_1) = \mathfrak{a}_1$, then we would have $\epsilon_1 = \bar{x}\epsilon_1\bar{x}^{-1}$; but it is clear that $\bar{C} = \bar{C}_+ + \bar{C}_+\bar{x}$; since ϵ_1 is in the center of $\bar{C}_+$, it would be in the center of $\bar{C}$. But this is impossible, since, $\bar{C}_+$ not being simple, $\bar{M}$ is of even dimension and $\bar{C}$ central simple. It follows that $j(\mathfrak{a}_1) = \mathfrak{a}_2$, $j(\mathfrak{a}_2) = \mathfrak{a}_1$. Since $\bar{M} \neq M$, the conjugate space of $\bar{M}$ contains some nonisotropic vector y; since y anticommutes with every element of $\bar{M}$, it commutes with every element of $\bar{C}_+$. The element $\bar{x}y$ is an invertible element of C_+ ; let j' be the mapping $u \to (\bar{x}y)u(\bar{x}y)^{-1}$ of C_+ into itself. Then j' extends j. Let $\mathfrak{M}_i$ be the set of elements $u \in C_+$ such that $\epsilon_i u = 0$ $(i = 1, 2)$; then it is clear that C_+ is the direct sum of $\mathfrak{M}_1$ and $\mathfrak{M}_2$ (because $\epsilon_1 + \epsilon_2 = 1$, $\epsilon_1\epsilon_2 = 0$). Since $j' \cdot \epsilon_1 = \epsilon_2$, $j' \cdot \epsilon_2 = \epsilon_1$, j' transforms $\mathfrak{M}_1$ into $\mathfrak{M}_2$, and $\mathfrak{M}_1$, $\mathfrak{M}_2$ have the same dimension. It is clear that $\bar{u}\mathfrak{M}_i \subset \mathfrak{M}_i$ $(i = 1, 2)$ for any $\bar{u} \in \bar{C}_+$; denote by $\theta_i(\bar{u})$ the restriction of $\theta(\bar{u})$ to $\mathfrak{M}_i$. Then θ is the sum of the representations θ_1 and θ_2 , which are of the same degree. One of the representations ω_1 , ω_2 map ϵ_1 upon 0 and ϵ_2 upon the identity, and the other maps ϵ_2 upon 0 and ϵ_1 upon the identity; we may assume that $\omega_i(\epsilon_i) = 0$. It is then clear that θ_i is the sum of a certain number of representations equivalent to ω_i $(i = 1, 2)$. Since θ_1 , θ_2 are of the same degree and ω_1 , ω_2 of the same degree, it is clear that ω_1 occurs as many times in θ_1 as ω_2 in θ_2 , and therefore that ω_1 and ω_2 occur the same number of times in θ and in τ. This shows that τ is the sum of a certain number of representations equivalent to the spin representation of $\bar{C}_+$.

Assume now that C_+ is not simple. Let ρ_1 be a half-spin representation of C_+ and τ' the representation of $\bar{C}_+$ induced by ρ_1 . We shall see that ω_1 , ω_2 also occur the same number of times in τ'. The algebra C_+ is the sum of two simple ideals of which one, say $\mathfrak{a}$, is represented faithfully under ρ_1 , while the other one is mapped upon $\{0\}$. Let σ_1 be the representation of C_+ which assigns to every $u \in C_+$ the mapping $w_1 \to uw_1$ of $\mathfrak{a}$ into itself. Then σ_1 is the sum of a certain number of representations equivalent to ρ_1 , and the representation θ' of $\bar{C}_+$ induced by σ_1 is the sum of a certain number of representations equivalent to τ': it will be sufficient to prove that ω_1 and ω_2 occur the same number of times in θ'. The automorphism j' defined above is an inner automorphism of C_+ and therefore transforms $\mathfrak{a}$ into itself. The proof then goes exactly as above, decomposing $\mathfrak{a}$ into the direct sum $\mathfrak{M}'_1 + \mathfrak{M}'_2$ of the spaces $\mathfrak{M}'_i = \mathfrak{M}_i \cap \mathfrak{a}$, which are transformed into each other by j'.

2.7. Extension of the Basic Field

Let M be a finite-dimensional vector space of even dimension over a field K, and Q a quadratic form on M whose associated bilinear form is nondegenerate. Let K' be an overfield of K, let M' be the vector space over K' which is deduced from M by extending to K' the basic field, and let Q' be the quadratic form on M' which extends Q. Then the Clifford algebra C' of Q' may be identified to the algebra deduced from C by extending the basic field to K' (II.1.5). It is clear that $C_+ = C'_+ \cap C$ and that the main antiautomorphism of C is the restriction to C of the main antiautomorphism of C'. Let Γ, Γ^+, $\Gamma_0{}^+$ be the Clifford group, the special Clifford group, and the reduced Clifford group of Q, and Γ', Γ'^{+}, $\Gamma'_0{}^{+}$ those of Q'. Then it is clear that

$$\Gamma \subset \Gamma' \qquad \Gamma^{+} \subset \Gamma'^{+} \qquad \Gamma_0{}^{+} \subset \Gamma_0'^{+}.$$

Let χ, χ' be the vector representations of Γ, Γ'. Then, for $s \in \Gamma$, $\chi(s)$ is obviously the restriction of $\chi'(s)$ to M. If M is even-dimensional, let ρ and ρ' be the spin representations of C and C'. If S is the space of spinors for Q, then ρ may be extended to a representation $\rho^{K'}$ of C' on the space $S^{K'}$ deduced from S by extension to K' of the basic field. Since $\rho^{K'}(1)$ is the identity, $\rho^{K'}$ is the sum of a certain number of representations of C' equivalent to the spin representation. It follows that, for any one of the groups Γ, Γ^{+}, $\Gamma_0{}^{+}$, the representation deduced from the spin representation by extension of the basic field is the sum of a certain number of representations all equivalent to the one induced by the spin representation of the corresponding group Γ', Γ'^{+}, or $\Gamma'_0{}^{+}$. If C_{+} is not simple, then the same is true of C'_{+} and the simple ideals of C'_{+} are those generated by the simple ideals of C_{+}. This shows that the representation of Γ^{+} or $\Gamma^{+}{}_0$ deduced by extension of the basic field of a half-spin representation is the sum of a certain number of representations all equivalent to the one induced by a suitable half-spin representation of the corresponding group Γ'^{+} or $\Gamma'_0{}^{+}$.

If M is odd-dimensional, we see in the same way that the representation of Γ^{+} or $\Gamma_0{}^{+}$ deduced by extending the basic field from the spin representation is the sum of a certain number of representations of all equivalent to the one induced by the spin representation of Γ'^{+} or $\Gamma'_0{}^{+}$.

2.8. The Theorem of Hurwitz

Let M be a vector space of finite dimension m over a field K and Q a quadratic form on M whose associated bilinear form B is nondegenerate. In certain cases, it is possible to find a bilinear mapping φ of $M \times M$ into M which satisfies the identity

$$Q(\varphi(x, y)) = Q(x)Q(y) \qquad (x, y \in M). \tag{1}$$

For instance, if $m = 1$ and Q takes the value 1, let x_1 be such that $Q(x_1) = 1$. If $x, y \in M$, set $x = ax_1$, $y = bx_1$; then the mapping defined by $\varphi(x, y) = abx_1$ has the required property.

Now, assuming that K is not of characteristic 2, let Z be a commutative algebra of dimension 2 over K with a base (x_1, x_2) such that x_1 is the unit element and $x_2^2 = ax_1$, a being an element $\neq 0$ in K. Then there is an automorphism $z \to \bar{z}$ of order 2 of Z such that $\bar{x}_2 = -x_2$;

if $x = ux_1 + vx_2 \in Z$ (with $u, v \in K$), then we have $x\bar{x} = (u^2 - av^2)x_1$; set $Q(x) = u^2 - av^2 = x\bar{x}$. Then the associated bilinear form of Q is clearly nondegenerate. We have

$$Q(xy)x_1 = xy\bar{x}\bar{y} = x\bar{x}y\bar{y} = Q(x)Q(y)x_1 ,$$

whence $Q(xy) = Q(x)Q(y)$.

The algebra Z defined above may be imbedded into a "generalized quaternion algebra L" which is generated by Z and by an element x_3 such that $x_3^2 = bx_1$, $b \in K$, $b \neq 0$, and $x_3x_2x_3^{-1} = -x_2 = \bar{x}_2$, whence $x_3zx_3^{-1} = \bar{z}$ for every $z \in Z$. The elements x_1, x_2, x_3, $x_2x_3 = x_4$ form a base of L. If

$$x = \sum_{i=1}^{4} u_ix_i \in L,$$

set

$$\bar{x} = u_1x_1 - \sum_{i=2}^{4} u_ix_i .$$

We then have $\bar{x}_1 = x_1$, $\bar{x}_i = -x_i$ if $i > 1$. The multiplication in L is defined by the formulas

$$x_1x_i = x_ix_1 \quad (i = 1, 2, 3, 4) \qquad x_2^2 = ax_1 \qquad x_3^2 = bx_1 \qquad x_4^2 = -abx_1 ,$$

$$x_2x_3 = -x_3x_2 = x_4 \qquad x_2x_4 = ax_3 = -x_4x_2 \qquad x_4x_3 = bx_2 = -x_3x_4 .$$

It follows immediately that the mapping $x \to \bar{x}$ is an antiautomorphism of L. The only elements left fixed by this antiautomorphism are those of Kx_1. The conjugate of $x\bar{x}$ being $x\bar{x}$ itself, we have $x\bar{x} = Q(x)x_1$, $Q(x)$ a scalar. An easy computation gives

$$Q\left(\sum_{i=1}^{4} u_ix_i\right) = u_1^2 - au_2^2 - bu_3^2 + abu_4^2, \qquad Q(xy) = Q(x)Q(y).$$

Were K of characteristic 2, we could construct similar examples by taking for Z a commutative algebra of dimension 2 over K which is either the sum of two fields isomorphic to K or a separable quadratic extension of K.

On the other hand, if M is of dimension 8 and Q of index 4, we shall construct in $V \cdot 5$ a mapping φ which has the property (1).

We shall now prove a result due to Hurwitz,[1] which states that, if

[1]A. Hurwitz, "Über die Komposition der quadritischen Formen von beliebig vielen Variabeln," *Nachrichten von der Königlichen Gesellschaft der Wissenschaften zu Göttingen*, 1898, p. 309, or *Mathematische Werke*, (Basel: Birkhauser, 1932), II, p. 565; see also "Über die Komposition der quadratischen Formen," *Mathematische Annalen*, *88* (1923), p. 1, or *Mathematische Werke*, II, p. 641.

$m \neq 1, 2, 4, 8$, there exists no bilinear mapping φ of $M \times M$ into M for which (1) holds.

Assume that we have a bilinear mapping φ for which (1) is true. Let x, y, z be elements of M; we then have $\varphi(x, y + z) = \varphi(x, y) + \varphi(x, z)$; applying (1) by replacing successively y by y, z and $y + z$, we easily obtain the formula

$$B(\varphi(x, y), \varphi(x, z)) = Q(x)B(y, z). \tag{2}$$

Replacing x by x, x' and $x + x'$ in (2), we easily obtain

$$B(\varphi(x, y), \varphi(x', z)) + B(\varphi(x', y), \varphi(x, z)) = B(x, x')B(y, z). \tag{3}$$

Conversely, assume that we have a subset S of M such that the formulas (1), (2), (3) are valid whenever x, x', y, z are in S. If S spans the vector space M, then (1) is valid for all $x, y \in M$. For, let first y be in S and x any element of M; write

$$x = \sum_{i=1}^{h} a_i x_i$$

with $a_i \in K$, $x_i \in S$. Then we have

$$\begin{aligned} Q(\varphi(x, y)) &= Q\left(\sum_{i=1}^{h} a_i \varphi(x_i, y)\right) \\ &= \sum_{i=1}^{h} a_i^2 Q(\varphi(x_i, y)) + \sum_{i<j} a_i a_j B(\varphi(x_i, y), \varphi(x_j, y)) \\ &= \sum_{i=1}^{h} a_i^2 Q(x_i)Q(y) + \sum_{i<j} a_i a_j B(x_i, x_j)Q(y) \\ &= Q(x)Q(y), \end{aligned}$$

which shows that (1) is true if $x \in M$, $y \in S$. On the other hand, if y, $z \in S$, we have, by using (2), (3),

$$\begin{aligned} B(\varphi(x, y), \varphi(x, z)) &= \sum_{i,j=1}^{h} a_i a_j B(\varphi(x_i, y), \varphi(x_j, z) \\ &= \sum_{i=1}^{h} a_i^2 Q(x_i)B(y, z) + \sum_{i<j} a_i a_j B(x_i, x_j)B(y, z) \\ &= Q(x)B(y, z), \end{aligned}$$

which shows that (2) is true if $x \in M$, $y, z \in S$. Now let x and y be arbitrary in M; writing y as a linear combination of elements of S, we see by a computation similar to the one made above that $Q(\varphi(x, y)) = Q(x)Q(y)$.

This being said, let K' be an overfield of K. Denote by M' the vector space over K' deduced from M by extending the basic field and by Q' the quadratic form on M' which extends Q. It is easily seen that φ may be extended to a bilinear mapping φ' of $M' \times M'$ into M'. Since M spans M', it follows from what we have just said that $Q'(\varphi'(x, y)) = Q'(x)Q'(y)$ for all $x, y \in M'$. This shows that, in proving Hurwitz's Theorem, we may replace K by any larger field. In particular, we may assume that Q has its maximal possible index r (i.e., $m = 2r$ or $m = 2r + 1$). We shall also assume that $m > 2$. The space M contains at least one element x_1 such that $Q(x_1) = 1$, for, if x, y are such that $Q(x) = Q(y) = 0$, $B(x, y) = 1$, then $x_1 = x + y$ has the required property. The mapping $\sigma_1 : y \to \varphi(x_1, y)$ then belongs to the orthogonal group G of Q. Set $\psi(x, y) = \sigma_1^{-1}. (\varphi(x, y))$. Then it is clear that $Q(\psi(x, y)) = Q(x)Q(y)$ and $\psi(x_1, y) = y$ for all $y \in M$. For any $x \in M$, we denote by $L(x)$ the mapping $y \to \psi(x, y)$. We first prove that m is even. Let x be any singular vector; then we have $Q(L_x \cdot y) = 0$ for all $y \in M$, and $L_x(M)$ is totally singular. The dimension ρ of this space is therefore $\leq r$. On the other hand, if y is a nonsingular vector, then the mapping $z \to \psi(z, y)$ is one-to-one. For, we easily see (in the same way that we proved formula (2)) that $B(\psi(z, y), \psi(z', y)) = B(z, z')Q(y)$ for all z, $z' \in M$. Thus, the condition $\psi(z', y) = 0$ implies $B(z, z') = 0$ for all $z \in M$, whence $z' = 0$. This shows that the kernel of L_x is totally singular. The dimension of this kernel is $m - \rho$; thus, we have $m \leq \rho + r = 2r$, which shows that $m \neq 2r + 1$ and that m is even.

Let x be any element of M such that $B(x, x_1) = 0$. Then $Q(\psi(x_1 + x, y)$ is equal on the one hand to

$$Q(x_1 + x)Q(y) = (1 + Q(x))Q(y)$$

and on the other hand to

$$Q(y + L_x \cdot y) = Q(y) + Q(x)Q(y) + B(y, L_x \cdot y);$$

thus, we have $B(y, L_x \cdot y) = 0$. Replacing y by y, z and $y + z$ in this formula, we obtain immediately

$$B(y, L_x \cdot z) + B(z, L_x \cdot y) = 0.$$

Replacing z by $L_x \cdot z$, we obtain

$$B(y, L_x^2 \cdot z) = -Q(x)B(y, z),$$

or

$$B(y, (L_x^2 + Q(x)I) \cdot z) = 0,$$

where I is the identity mapping. We conclude that

$$L_x^2 = -Q(x) \cdot I \qquad \text{if} \qquad B(x, x_1) = 0. \tag{4}$$

Let H be the conjugate hyperplane of Kx_1. If K is of characteristic 2, let N be a subspace of H supplementary to Kx_1; if not, let N be any $(m - 2)$-dimensional nonisotropic subspace of H. Let C' be the Clifford algebra of the restriction of $-Q$ to N, and let $\mathfrak{M}$ be the algebra of all endomorphisms of M. It follows from (4) and from II.1.1 that the linear mapping $x \to L_x$ of N into $\mathfrak{M}$ may be extended to a homomorphism θ of C' into $\mathfrak{M}$. The space N is not isotropic; since N is of even dimension $m - 2$, C' is central simple and θ is an isomorphism. The algebra C' is of dimension $2^{2(r-1)}$; its simple representations are therefore all equivalent, and their degrees are multiples of 2^{r-1}. Since $\theta(C')$ contains the identity, $m = 2r$ must be a multiple of 2^{r-1}, and r is a multiple of 2^{r-2}. It is easily seen that this can happen only for $r = 2$ or 4 (since $m > 2$ by assumption), which proves Hurwitz's Theorem.

2.9. Quadratic Forms over the Real Numbers

Let M be a vector space of finite dimension m over the field of real numbers, and Q a quadratic form of rank m over M. If x is not singular in M, then there is a real number a such that $Q(ax) = \pm 1$. Thus, M has a base $(x_1, \cdots, x_m)$ composed of mutually orthogonal vectors x_i such that $Q(x_i) = \pm 1$. We may assume that $Q(x_{2k-1}) = 1$, $Q(x_{2k}) = -1$ for $1 \leq k \leq \nu$, ν being some integer $\leq m/2$, while the $Q(x_i)$ are all equal to each other for $i > 2\nu$; let ϵ be their common value. Then we have

$$Q\left(\sum_{i=1}^{m} a_i x_i\right) = \sum_{k=1}^{\nu} (a_{2k-1}^2 - a_{2k}^2) + \epsilon \sum_{i=2\nu+1}^{m} a_i^2,$$

where $\epsilon = \pm 1$. If we denote by N the space spanned by the elements $x_{2k-1} + x_{2k}$ $(1 \leq k \leq \nu)$, by P the space spanned by the elements $x_{2k-1} - x_{2k}$ $(1 \leq k \leq \nu)$, and by R the space spanned by $x_{2\nu+1}, \cdots, x_m$, then $M = N + P + R$, N, and P are totally singular and the restriction of Q to R is definite (positive if $\epsilon > 0$, negative if $\epsilon < 0$). The conjugate of N is $N + R$, and the only singular vectors of this space are those of N, which shows that ν is the index of Q. If we denote by p the number of indices i such that $Q(x_i) = 1$ and by q the number of indices i for which $Q(x_i) < 0$, then ν is the smallest of p, q, while $p + q = m$; ϵ is $+1$ if $p > q$, -1 if $p < q$. Let $(x'_1, \cdots, x'_m)$ be any other base with the same properties as $(x_1, \cdots, x_m)$, and let N', P', R', p', q', ϵ' be determined for this new base as N, P, R, p, q, ϵ have been for $(x_1, \cdots, x_m)$. The restrictions of Q to $N + P$, $N' + P'$ are equivalent; the same is therefore true of its restrictions to R, R', whence $\epsilon = \epsilon'$ (if $2\nu \neq m$). We have $p' + q' = p + q$, $\min\{p, q\} = \min\{p', q'\}$, and $p > q$ if and only if $p' > q'$; it follows that $p = p'$, $q = q'$. This is the famous *law of inertia*.

Let us now determine the structure of the Clifford algebra C of Q. The discriminant of the restriction of B to $(N + P) \times (N \times P)$ is $(-1)^r 2^{2r}$. Thus, C is isomorphic to the tensor product of the Clifford algebras C_0, C_1 of the restrictions of Q to $N + P$ and R (II.2.5), and C_0 is isomorphic to a full matrix algebra (II.2.1). Let us now assume that Q is of index 0, and suppose first that $m = 2r$ is even. There is only one central division algebra $\neq K$ over the field K of real numbers (up to isomorphism), and this is the algebra Ω of quaternions. Thus, C is either isomorphic to a full matrix algebra over K or to a full matrix algebra over Ω. Set $\zeta(Q) = +1$ in the first case, $\zeta(Q) = -1$ in the second case. Also set $\epsilon(Q) = +1$ or -1, according to whether Q is positive or negative definite. If $m = 2$, then C has a base $(1, x_1, x_2, x_1x_2)$ such that $x_1^2 = \epsilon\cdot 1$, $x_2^2 = \epsilon\cdot 1$, $(x_1x_2)^2 = -1$. If $\epsilon = +1$, then C has zero divisors (for instance, $x_1 - 1$); if $\epsilon = -1$, then we recognize the classical base of Ω; thus, $\zeta(Q) = \epsilon(Q)$ if $m = 2$. If $m > 2$, let N be a nonisotropic subspace of dimension $m - 2$ of M, and N' the conjugate of N; then C is isomorphic to the tensor product of the Clifford algebra of the restriction Q_N of Q to N by that of $(-1)^{r-1}Q_{N'}$, where $Q_{N'}$ is the restriction of Q to N' (II.2.5, observing that the discriminant of the restriction of B to $N \times N$ is a square). Since $\Omega \otimes \Omega$ is a full matrix algebra, we have

$$\zeta(Q) = \zeta(Q_N)\zeta((-1)^{r-1}Q_{N'}) = (-1)^{r-1}\zeta(Q_N)\epsilon.$$

It follows immediately that

$$\zeta(Q) = (-1)^{r(r-1)/2}\epsilon^r(Q).$$

Moreover, C^+ is simple if r is odd, and is the direct sum of two ideals if r is even (by II.2.3).

Suppose now that $m = 2r + 1$ is odd. Let x_0 be any element $\neq 0$ in M, and N the conjugate of Kx_0. Then C_+ is isomorphic to the Clifford algebra of the restriction of $-\epsilon Q$ to N (see the proof of II.2.6). Thus, C_+ isomorphic to a full matrix algebra over K if $(-1)^{r(r+1)/2} = 1$, over Ω if $(-1)^{r(r+1)/2} = -1$. Moreover, C is simple if $(-1)^r\epsilon = -1$, but is the sum of two simple ideals if $(-1)^r\epsilon = +1$.

Returning now to the general case, we observe that the Clifford group Γ of Q is a closed subgroup of the multiplicative group of invertible elements of C (where C is given its natural vector-space topology); Γ is therefore a Lie group. We shall determine its Lie algebra. For any $X \in C$, let $L(X)$ be the operator of left multiplication by X in C. Then $L(X)$ is an endomorphism of the finite-dimensional vector space C; as such, it has an exponential

$$\exp L(X) = \sum_{k=0}^{\infty} (k!)^{-1}(L(X))^k.$$

We may write this as

$$\lim_{n \to \infty} L\Big(\sum_{k=0}^{n} (k!)^{-1}X^k\Big).$$

It is clear that $u \to L(u)$ is a homeomorphism of C with a subspace of the vector space of endomorphisms of the vector space structure of C. We conclude that

$$\sum_{k=0}^{n} (k!)^{-1}X^k$$

tends to a limit in C as n increases indefinitely. We denote this limit by $\exp X$, and we then have $\exp L(X) = L(\exp X)$. We know that the exponential of a matrix depends continuously on this matrix; it follows that $X \to \exp X$ is a continuous mapping of C into itself. We have $\exp (X + Y) = (\exp X)(\exp Y)$ if $XY = YX$; in particular, $\exp X$ is invertible, and $(\exp X)^{-1} = \exp (-X)$. Let C^* be the multiplicative group of invertible elements of C. Then $t \to \exp tX$ ($t \in K$) is a one-parameter subgroup of C^*; thus, we see that $L(C)$ is in the Lie algebra of $L(C^*)$. But $L(C^*)$ is obviously of dimension $\leq 2^m$ and $L(C)$ is of dimension 2^m. Thus, $L(C)$ is the full Lie algebra of $L(C^*)$. If we set $[X, Y] = XY - YX$ for X, Y in C, we have $L([X, Y]) = [L(X), L(Y)]$; thus, we see that we may regard C as the Lie algebra of C^*, C being made into a Lie algebra by means of the law of composition $(X, Y) \to [X, Y]$.

If $u \in C^*$, denote by $\chi(u)$ the mapping $w \to uwu^{-1}$ of C into itself; χ is a linear representation of C^*. Regarding C as the Lie algebra of C^*, χ is clearly the adjoint representation of C^*. If we denote by $A(X)$ the mapping $Y \to [X, Y]$, then $\chi(\exp X) = \exp A(X)$. Now, Γ is the group of all $u \in C^*$ such that $(\chi(u))(M) = M$; for X to belong to the Lie algebra of Γ, it is necessary and sufficient that $\exp tA(X)$ should map M into itself for all real t, i.e., that $A(X)$ should map M into itself. We propose now to determine the elements X with this property.

Let x, y be in M; then we have, for $z \in M$,

$$\begin{aligned} xyz - zxy &= xB(y, z) - (xz + zx)y \\ &= B(y, z)x - B(x, z)y \end{aligned}$$

and xy belongs to the Lie algebra of Γ. If m is odd, then the center Z of C is spanned by 1 and by an odd element z, and $A(Z) = 0$. Let c be the space spanned by the elements xy ($x, y \in M$) and by z (the last one only if m is odd). Then $\mathfrak{c}$ is clearly of dimension $1 + m(m - 1)/2$ if m is even, $2 + m\,(m - 1)/2$ if m is odd. Now, the image of Γ under its vector representation is G or G^+ (depending on the parity of m), and it is well known that G and G^+ are of dimension $m(m - 1)/2$. The representation χ is continuous, and its kernel is the intersection of Γ with the center of Z; this kernel is of dimension 1 if m is even, 2 if m is odd. This shows that Γ is of dimension $1 + m(m - 1)/2$ if m is even, $2 + m(m - 1)/2$ if m is odd. This shows that $\mathfrak{c}$ *is the full Lie algebra of* Γ. The Lie algebra of Γ^+ is obviously the space $\mathfrak{c}^+$ spanned by the products xy, $x, y \in M$. If α is the main antiautomorphism of C, then, clearly, $\alpha\,(\exp X) = \exp \alpha(X)$ ($X \in C$). It follows immediately that the Lie algebra $\mathfrak{c}_0^+$ of Γ_0^+ is the set of $X \in \mathfrak{c}^+$ such that $\alpha(X) + X = 0$. This is easily seen to be the space spanned by all products xy, where x, y are vectors of M orthogonal to each other.

If Q is definite (either positive or negative), then it is well known that G^+ is a connected group. In that case, we have $G_0^+ = G^+$ in virtue of II.3.6. The kernel of the vector representation χ of Γ_0^+ is composed of 1 and -1. If $m > 1$, then -1 belongs to the connected component of 1 in Γ_0^+. For, let x and y be two vectors of M orthogonal to each other, such that $Q(x) = Q(y) = \pm 1$. Then we have $(xy)^2 = -1$ and $\exp txy = \cos t + (\sin t)xy$, whence $\exp \pi xy = -1$; since $\exp txy \in \Gamma_0^+$ for all real t, -1 belongs to the connected component of 1. It follows easily that Γ_0^+ is a connected group, which "covers" G^+ exactly twice. If $m > 1$, then it is known that the Poincaré group of G^+ is of order 2; Γ_0^+ is then the simply connected covering group of G^+.

If, however, Q is of index $\neq 0$, then G_0^+ is of index 2 in G^+ (by II.3.6). Every element σ sufficiently near the identity in the Lie group G^+ belongs to a one-parameter subgroup and is therefore a square in G^+, whence $\sigma \in G_0^+$. It follows that G_0^+ is an open subgroup of G^+ and contains the connected component of the identity in G^+. We shall establish that G_0^+ is connected. If x is a nonsingular vector in M, then there is a scalar a such that $Q(ax) = \pm 1$. Thus, an element $s \in \Gamma^+$ may be represented in the form $s = cx_1 \cdots x_{2h}$, where $x_i \in M$, $Q(x_i) = \pm 1$ ($1 \leq i \leq 2_h$), and $c \in K$. If p is the number of indices i such that $Q(x_i) = -1$, then we have $\lambda(s) = (-1)^p c^2$; thus, s belongs to Γ_0^+ if and only if $c = \pm 1$ and p is even. Moreover, we may assume that $Q(x_1) = \cdots = Q(x_p) = -1$, $Q(x_{p+1}) = \cdots = Q(x_{2h}) = +1$. For, if $Q(x_i) = 1$, $Q(x_{i+1}) = -1$, we may write

$$x_i x_{i+1} = x_{i+1}(x_{i+1}^{-1}\, x_i x_{i+1})$$

and $Q(x_{i+1}^{-1}x_ix_{i+1}) = +1$; by a succession of transformations of this kind, we may bring all factors x_i with $Q(x_i) < 0$ in front of the product. Thus, Γ_0^+ is generated by ± 1 and by the products xy, where x, y are vectors such that $Q(x) = Q(y) = \pm 1$. Consider now any such product xy. If x, y are linearly dependent, then we have $xy = \pm 1$. If not, then they span a plane P. Let D be the set of vectors $x' \in P$ such that $Q(x') = Q(x)$. We shall see that y belongs either to the connected component of x or to that of $-x$ in P. If P is isotropic, then $P = Kx + Kz$, where z is a singular vector such that $B(x, z) = 0$, and D consists of all vectors $\pm x + az$, $a \in K$, which proves our assertion in that case. If P is not isotropic and the restriction of Q to P is of index 1, then we have $P = Kz + Kz'$, where z, z' are singular vectors such that $B(z, z') = 1$. In that case, D is the set of all vectors of the form $kz + Q(x)k^{-1}z'$, with $k \neq 0$; D has two components (corresponding to the cases where $k > 0$, $k < 0$), one of which contains x and the other $-x$. If the restriction of Q to P is of index 0, then D is clearly connected. It follows that xy belongs either to the component of 1 or to that of -1 in Γ_0^+. Thus, Γ_0^+ has at most two connected components, and, if it has two, then one of them contains 1 and the other -1. It follows immediately that $G_0^+ = \chi(\Gamma_0^+)$ is connected. It is easily seen that Γ_0^+ itself is connected if $m > 2$, but not if $m = 2$.

CHAPTER III

FORMS OF MAXIMAL INDEX

We shall denote by M a vector space of finite dimension m over a field K and by Q a quadratic form on M whose associated bilinear form B is nondegenerate. We shall furthermore assume that Q is of maximal index, i.e., of index $m/2$ if m is even, $(m - 1)/2$ if m is odd. We shall denote by G the orthogonal group of Q, by G^+ its group of rotations, by G_0^+ its restricted orthogonal group, by C its Clifford algebra, by C_+ and C_- the spaces of even and odd elements of C, by Γ the Clifford group of Q, by χ the vector representation of Γ, by Γ^+ the special Clifford group of Q, by Γ_0^+ its reduced Clifford group, by ρ, ρ^+, ρ_0^+ the spin representations of Γ, Γ^+, Γ_0^+ (the first one only in the case m even), by λ the norm homomorphism, and by α the main antiautomorphism of C.

Except in Section 3.8, we shall assume that m is even and we shall set $m = 2r$. We shall then denote by N and P fixed totally singular r-dimensional subspaces of M such that $M = N + P$, by C^N and C^P the subalgebras of C generated by N and P, and by f the product of the elements of some base of P. Then Cf is a minimal left ideal of C, and we have $Cf = C^N f$ (II.2.2). There is a representation ρ of C on C^N such that $vuf = (\rho(v) \cdot u)f$ if $v \in C$, $u \in C^N$. Since Cf is a minimal left ideal, ρ is simple. We may therefore take the space S of spinors to be C^N, ρ being the spin representation. We shall always assume that S has been defined in this manner.

The space C^N may be identified to the exterior algebra of N. For any integer h, let C_h^N be the space of homogeneous elements of degree h of C^N; $C^N \cap C_+$ is then the sum of the spaces C_h^N for h even, while $C^N \cap C_-$ is the sum of the spaces C_h^N for h odd; we shall denote these spaces by C_+^N, C_-^N.

If $x \in N$, then $\rho(x)$ is the operation of left mutiplication by x in C^N, while, if $y \in P$, $\rho(y)$ is the homogeneous antiderivation of degree -1 of C^N such that $\rho(y) \cdot x = B(x, y) \cdot 1$ for $x \in N$ (II.2.2). It follows immediately that, if $z \in M$, then $\rho(z)$ maps C_+^N into C_-^N and C_-^N into C_+^N. We conclude that, if $u \in C_+$, then $\rho(u)$ maps each one of the spaces C_+^N, C_-^N into itself; we denote by $\rho_p^+(u)$, $\rho_i^+(u)$ the restrictions of $\rho(u) =$

$\rho^+(u)$ to $C_+{}^N$, $C_-{}^N$. Thus, we see that the spin representation ρ^+ of C_+ is not simple. We shall deduce from this that C_+ itself is not simple. The algebra C_+ is of dimension 2^{m-1}, while the algebra of all endomorphisms of the vector space C^N is of dimension 2^m. Since C_+ is semi-simple, the algebra $\mathfrak{Z}$ of vector-space endomorphisms of C^N which commute with all operations of $\rho(C_+)$ is of dimension $2^m/2^{m-1} = 2$. The center Z of C_+ is of dimension 2 (II.2.3), and $\mathfrak{Z} \supset \rho(Z)$, whence $\mathfrak{Z} = \rho(Z)$. Were C_+ simple, then Z would be a field; the same would be true of $\mathfrak{Z}$, and, by a well-known theorem, ρ^+ would be simple, which is not the case.

Thus, the spaces of half-spinors are $C_+{}^N$ and $C_-{}^N$; we shall denote $C_+{}^N$ (respectively: $C_-{}^N$) by S_p (respectively: S_i) and call it the space of *even* (respectively: *odd*) *half-spinors*. The half-spin representations of Γ^+, $\Gamma_0{}^+$ on the spaces S_p, S_i will be denoted by $\rho_p{}^+$, $\rho_i{}^+$ for Γ^+, and by $\rho_{0,p}{}^+$, $\rho_{0,i}{}^+$ for $\Gamma_0{}^+$.

3.1. Pure Spinors

Let Z be any totally singular subspace of dimension r of M, C^Z the subalgebra of C generated by Z, and f_Z the product of the elements of some base of Z. Then f_Z is determined by Z up to a scalar factor $\neq 0$ (as follows from the fact that C^Z may be identified to the exterior algebra of Z), and $f^Z C$ is a minimal right ideal of C (by II.2.2).

III.1.1. *The intersection of any minimal left ideal of C with any minimal right ideal is a vector space of dimension* 1 *over* K.

The algebra $\rho(C)$ is the algebra of all endomorphisms of S. Let $\mathfrak{a}$ be a minimal left ideal of C: then we have $\mathfrak{a} = Ce$, where e is an idempotent. The operation $\rho(e)$, being idempotent, is a projection; let H be its kernel and $H' = (\rho(e))(S)$. It is clear that for any $v \in C$, $\rho(ve)$ maps H upon $\{0\}$. Conversely, let v' be in C and such that $\rho(v')$ maps H upon $\{0\}$; it is then clear that $v' = v'e$ (since $\rho(e)$ maps the elements of H upon themselves). Thus, $\mathfrak{a}$ is the set of all elements $v \in C$ such that $\rho(v)$ maps H upon $\{0\}$. Conversely, if H_1 is any subspace of S, the set $\mathfrak{a}_1$ of elements $v \in C$ such that $\rho(v)$ maps H_1 upon $\{0\}$ is a left ideal, and, if $H_1 \supset H$, then $\mathfrak{a}_1 \subset \mathfrak{a}$. Since $\mathfrak{a}$ is minimal, it follows immediately that H is a hyperplane. Let now $\mathfrak{b}$ be a minimal right ideal; then $\mathfrak{b} = e'C$, where e' is an idempotent. The operation $\rho(e')$ is a projection; let D be its kernel and $D' = (\rho(e'))(S)$; then the operations of $\mathfrak{b}$ map S into D'. Conversely, let v be in C and such that $(\rho(v))(S) \subset D'$. Since $\rho(e')$ is the identity on D', we have $v = e'v \in \mathfrak{b}$. Conversely, for any subspace D_1' of S, the set of $v \in C$ such that $\rho(v)$ maps S into D_1' is a right ideal

$\mathfrak{b}_1$, and $D_1' \subset D'$ implies $\mathfrak{b}_1 \subset \mathfrak{b}$. Since $\mathfrak{b}$ is minimal, D' must be of dimension 1. Now, $\mathfrak{a} \cap \mathfrak{b}$ is the set of $v \in C$ such that $\rho(v)$ maps H upon $\{0\}$ and S into D'; if x is in S but not in H and $y \neq 0$ in D', then $v \in \mathfrak{a} \cap \mathfrak{b}$ implies that $\rho(v) \cdot x = ay$, $a \in K$, and v is uniquely determined when a is given. It follows that $\mathfrak{a} \cap \mathfrak{b}$ is of dimension 1.

This being said, let us return to the notation used above. The space $Cf \cap f_Z C$ is one-dimensional, and may therefore be written in the form $S_Z f$, where S_Z is a one-dimensional subspace of S. Any element $\neq 0$ of this space is called a *representative spinor* of Z. Any element of S which is representative of some r-dimensional totally singular space is called a *pure spinor*.

III.1.2. *Let Z be a totally singular r-dimensional subspace of M. Then there exists an $s \in \Gamma$ such that $sPs^{-1} = Z$; for any such s, $\rho(s) \cdot 1$ is a representative spinor for Z.*

Any vector-space isomorphism of P with Z may be extended to an operation σ of G (by I.4.1); there is an $s \in \Gamma$ such that $\chi(s) = \sigma$ (by II.3.1). It is clear that $sPs^{-1} = Z$; it follows that $f_Z = sfs^{-1}$ is $\neq 0$ and is the product of the elements of a base of Z. We have $sf \in Cf$, $sf = f_Z s \in f_Z C$; thus, sf spans $Cf \cap f_Z C$ and $\rho(s) \cdot 1$ is a representative spinor for Z.

III.1.3. *Let Z be a totally singular r-dimensional subspace of M and u_Z a representative spinor for Z. If $s \in \Gamma$, then $\rho(s) \cdot u_Z$ is a representative element for the space sZs^{-1}.*

This follows immediately from III.1.2.

III.1.4. *Let Z be a totally singular r-dimensional subspace of M and u_Z a representative spinor for Z. Then Z is the set of elements $x \in M$ such that $\rho(x) \cdot u_Z = 0$. If $u \in S$ is such that $\rho(x) \cdot u = 0$ for all $x \in Z$, then $u = au_Z$ with some $a \in K$.*

If $u \in S$, $x \in M$, $s \in \Gamma$, the conditions $\rho(x)u = 0$, $\rho(s^{-1}xs) \cdot (\rho(s^{-1}) \cdot u) = 0$ are equivalent. It is therefore sufficient to prove the first assertion of III.1.4 in the case where $Z = P$; in that case, we may obviously assume that $u_Z = 1$. If $x' \in N$, $x'' \in P$, then $\rho(x') \cdot 1 = x'$ and $\rho(x'')$ is an anti-derivation of $C^N = S$, which maps 1 upon 0, whence $\rho(x' + x'') \cdot 1 = x'$. This is 0 if and only if $x' = 0$, i.e., $x \in P$. Since Z may be transformed into N by an operation of G, it is sufficient to prove the second assertion in the case where $Z = N$. Let then u be an element of C^N such that $\rho(x) \cdot u = 0$ for all $x \in N$; since C^N is the exterior algebra of N and $\rho(x) \cdot u = xu = x \wedge u$, it is well known that this implies that u is a scalar multiple of a basic element e of $C^N{}_r$; it follows that $Ku_N = Ke$, and III.1.4 is proved.

It follows from III.1.4 that a totally singular r-dimensional subspace of M is uniquely determined when any representative spinor of it is given.

III.1.5. *A representative spinor of any totally singular r-dimensional subspace of M is always a half-spinor (i.e., either even or odd). If $m > 0$, there are both even and odd pure spinors.*

We know that any $s \in \Gamma$ is either even or odd (III.3.2), and that, if $m > 0$, then there are even and odd elements in Γ; III.1.5 therefore follows from III.1.2.

We shall call *even* (respectively: *odd*) those maximal totally isotropic subspaces of M whose representative half-spinors are even (respectively: odd).

III.1.6. *Let Z and Z' be maximal totally singular subspaces of M. A necessary and sufficient condition for Z and Z' to be transformable into each other by an operation of G^+ is that Z and Z' be both even or both odd.*

We know that Z may be transformed into Z' by an operation of G, i.e., that there is an $s \in \Gamma$ such that $Z' = sZs^{-1}$; III.1.6 then follows immediately from III.1.2.

Let $(x_1, \cdots, x_r)$ be a base of N. If $u = \Sigma_{i<j} a_{ij}x_ix_j$ is an element of C_2^N, we set $\exp u = \Pi_{i<j}(1 + a_{ij}x_ix_j)$ (observe that the elements of C_2^N are in the center of C^N). Since $(x_ix_j)^2 = 0$, it is clear that

$$\exp(u + u') = (\exp u)(\exp u')$$

for any u, u' in C_2^N. If $u = ax_kx$, $x = \Sigma_{i=1}^r a_ix_i$, then we have $\exp u = 1 + u$, as follows immediately from the fact that $(x_kx_i)(x_kx_j) = 0$. Using the formula $\exp(u + u') = (\exp u)(\exp u')$ and observing that $(x_kx)(x_1x) = 0$, we see immediately that $\exp u = 1 + u$ whenever $u = yx$ is a decomposable element. This proves that our definition of $\exp u$ does not depend on the special base we have selected in N. We have $\exp 0 = 1$; it follows that $\exp u$ is always invertible and that $(\exp u)^{-1} = \exp(-u)$. For any $u \in C_2^N$, $\exp u$ is in C_+^N and differs from $1 + u$ by an element of $\Sigma_{k>1} C_{2k}^N$.

III.1.7. *If $u \in C_2^N$, then $\exp u$ belongs to Γ_0^+, and $\chi(\exp u)$ leaves the elements of N fixed. Any operation of G which leaves all elements of N fixed is in G_0^+ and may be written in the form $\chi(\exp u)$, $u \in C_2^N$. If Z, Z', Z'' are maximal totally singular subspaces of M such that $Z' \cap Z = Z'' \cap Z$, then there is an operation of G_0^+ which leaves all points of Z fixed and which transforms Z' into Z''.*

The elements $\exp u$, $u \in C_2^N$, obviously form a group H, image of the additive group of C_2^N under the homomorphism $u \to \exp u$. Every element of C_2^N being a sum of decomposable elements, in order to prove that $\exp u \in \Gamma$, it will be sufficient to show that $\exp (x_1x_2) \in \Gamma$ if x_1, x_2 are elements of N. We then have $x_2x_1 = -x_1x_2$, $x_1^2 = x_2^2 = 0$, and, for $y \in M$,

$$x_iy + yx_i = B(x_i, y)\cdot 1 \qquad (i = 1, 2),$$

whence, if $u = x_1x_2$,

$$\begin{aligned}(\exp u)y(\exp u)^{-1} &= (1 + x_1x_2)y(1 - x_1x_2)\\ &= y + B(x_2, y)x_1 - B(x_1, y)x_2 . \qquad (1)\end{aligned}$$

This shows that $\exp u \in \Gamma$. It is obvious that $\exp u \in \Gamma^+$. If $x_1, x_2 \in N$, then we have $\alpha(x_1x_2) = x_2x_1 = -x_1x_2$; thus, if u is decomposable, we have $\alpha(\exp u) = \exp(-u) = (\exp u)^{-1}$, whence $\exp u \in \Gamma_0^+$; the same is therefore true for all $u \in C_2^N$. Since $\exp u$ commutes with every element of N, $\chi(\exp u)$ leaves the elements of N fixed.

Coming back to formula (1), we observe that, if $y \in P$, then $\rho(y)$ is an antiderivation of C^N which maps any $x \in N$ upon $B(x, y)\cdot 1$, whence $\rho(y)\cdot x_1x_2 = B(x_1, y)x_2 - B(x_2, y)x_1$. Thus, (1) may be written as

$$\chi(\exp u)\cdot y = y - \rho(y)\cdot u \qquad (y \in P,\ u = x_1x_2).$$

Now, if $u = u_1 + \cdots + u_h$, each u_i being decomposable, then

$$\exp u = \prod_{i=1}^{h} (\exp u_i)$$

and $\chi(\exp u_i)$ leaves the elements of N fixed. It follows immediately that the formula

$$\chi(\exp u)\cdot y = y - \rho(y)\cdot u \qquad (y \in P)$$

is valid for every $u \in C_2^N$. Now, let σ be any operation of G which leaves all elements of N fixed. We have seen in the proof of I.4.5 that there exist bases $(x_1, \cdots, x_r)$ of N and $(y_1, \cdots, y_r)$ of P such that $B(x_i, y_j) = \delta_{ij}$ $(1 \le i, j \le r)$ and $\sigma\cdot y_{2k-1} = y_{2k-1} - x_{2k}$, $\sigma\cdot y_{2k} = y_{2k} + x_{2k-1}$ for $k \le \rho$, ρ being an integer $\le r/2$, while $\sigma\cdot y_i = y_i$ for $i > 2\rho$. If we set $u = x_1x_2 + \cdots + x_{2\rho-1}x_{2\rho}$, then $\sigma\cdot y = \chi(\exp u)\cdot y$ for $y \in P$, whence $\sigma = \chi(\exp u)$, since both sides leave the elements of N fixed and $M = N + P$.

In order to prove the last assertion, we first observe that there is a $\tau \in G$ such that $\tau(Z) = N$ (I.4.1). Set $Z'_1 = \tau(Z')$, $Z''_1 = \tau(Z'')$; then

Z'_1, Z''_1 are maximal totally singular spaces which have the same intersection with N. If there is a $\sigma_1 \varepsilon G_0^+$ which leaves the points of N fixed and transforms Z'_1 into Z''_1, then $\tau\sigma_1\tau^{-1}$ is in G_0^+ (for G_0^+ is a normal subgroup of G), transforms Z' into Z'' and leaves the elements of Z fixed. Thus, we see that we may assume that $Z = N$.

Let $(x_1, \cdots, x_h)$ be a base of $Z' \cap N$ and $(x_1, \cdots, x_r)$ a base of N containing $x_1, \cdots, x_h$. Let $(y_1, \cdots, y_r)$ be a base of P such that $B(x_i, y_j) = \delta_{ij}$ $(1 \leq i, j \leq r)$ and let P_1 be the subspace of P spanned by $y_{h+1}, \cdots, y_r$. It is clear that P_1 is the intersection of P with the conjugate of $Z' \cap N$, and that $Z_2' = (Z' \cap N) + P_1$ is a maximal totally singular subspace of M such that $Z'_2 \cap N = Z' \cap N$. It will be sufficient to prove that Z', Z'' may be transformed into Z'_2 by operations of G leaving the points of N fixed, for we know that these operations will then belong to G_0^+. It will furthermore be sufficient to present the argument in the case of Z'. We may represent Z' as the direct sum of $Z' \cap N$ and of a space U' of dimension $r - h$. If $z \varepsilon U'$, we write $z = f(z) + g(z)$, $f(z) \varepsilon N$, $g(z) \varepsilon P$. Since $U' \cap N = \{0\}$, g is a linear isomorphism of U' with a subspace of P. If $x \varepsilon Z' \cap N$, then we have $B(x, z) = 0$ and $B(x, f(z)) = 0$ because N and Z' are totally isotropic. It follows that $B(x, g(z)) = 0$, whence $g(z) \varepsilon P_1$; since $g(U')$ and P_1 are of dimension $r - h$, g is a linear isomorphism of U' with P_1. The sum $N + U'$ is direct; let $\bar{g}$ be the linear mapping of $N + U'$ into M which coincides with the identity on N and with g on U'; then $\bar{g}(Z') = Z_2'$. Moreover, $\bar{g}$ is a Q-isomorphism. For, let x be in N and z in U'; then $Q(\bar{g}(x + z)) = Q(x + g(z)) = B(x, g(z)) = B(x, z)$ because $B(x, f(z)) = 0$; but $Q(x + z)$ is also $B(x, z)$, since $Q(x) = Q(z) = 0$, which proves our assertion. Thus, $\bar{g}$ may be extended to an operation $\sigma \varepsilon G$ (I.4.1); σ leaves the elements of N fixed and maps Z' onto Z_2', which concludes the proof.

III.1.8. *Let $x_1, \cdots, x_h$ be linearly independent elements of N. Denote by A the space spanned by $x_1, \cdots, x_h$ and by A' the intersection of P with the conjugate of A. Then $Z' = A + A'$ is a maximal totally singular subspace of M, and $x_1 \cdots x_h$ is a representative spinor for Z'.*

It is clear that Z' is totally singular. We have $\dim A = h$ and $\dim A' = r - h$, since the restriction of B to $N \times P$ is nondegenerate; since $A \cap A' = \{0\}$, we have $\dim Z' = r$. If $x \varepsilon A$, then we have $\rho(x) \cdot x_1 \cdots x_h = x x_1 \cdots x_h = 0$. If $y \varepsilon A'$, then $\rho(y)$ is an antiderivation of C^N which maps x_i upon $B(x_i, y) \cdot 1 = 0$ $(1 \leq i \leq h)$, whence $\rho(y) \cdot x_1 \cdots x_h = 0$. Thus, we have $\rho(z) \cdot x_1 \cdots x_h = 0$ for all $z \varepsilon Z'$, which shows (by III.1.4) that $x_1 \cdots x_h$ is a representative spinor for Z'.

III.1.9. *A necessary and sufficient condition for a spinor u to be pure is that u be representable in the form $c(\exp v)x_1 \cdots x_h$, where $x_1, \cdots, x_h$ are linearly independent elements of N, $c \in K$, $c \neq 0$, and $v \in C_2^N$. If u is representative for the maximal totally singular space Z, then $x_1, \cdots, x_h$ form a base of $Z \cap N$.*

If $x_1, \cdots, x_h$ are linearly independent, then $x_1 \cdots x_h$ is pure and representative for a space Z_1 such that $Z_1 \cap N = Kx_1 + \cdots + Kx_h$ (III.1.8). If $v \in C_2^N$, then $\exp v \in \Gamma \cap C^N$, and $\rho(\exp v)$ is the operation of multiplication by $\exp v$ in C^N. Thus, $(\exp v)x_1 \cdots x_h$ is representative for the space $(\chi(\exp v))(Z_1)$, whose intersection with N is the same as that of Z_1, since $\chi(\exp v)$ leaves the elements of N fixed. Conversely, let Z be any maximal totally singular subspace of M, and $(x_1, \cdots, x_h)$ a base of $Z \cap N$. Define A, A', Z' as in III.1.8. Then it follows from III.1.7 that there is a $v \in C_2^N$ such that $\chi(\exp v)$ transforms Z' into Z; $(\exp v)x_1 \cdots x_h$ is therefore a representative spinor for Z.

III.1.10. *Let Z, Z' be maximal totally singular subspaces of M, and $h = \dim(Z \cap Z')$. If $h \equiv r \pmod 2$, then Z, Z' are of the same kind (both even or both odd); if not, then Z, Z' are of opposite kinds.*

There is an operation σ of G which transforms Z into N, and it is clear that Z and Z' are of the same kind if and only if $\sigma(Z)$ and $\sigma(Z')$ are (for, if $\tau \in G$ transforms Z into Z', then $\sigma\tau\sigma^{-1}$ transforms $\sigma(Z)$ into $\sigma(Z')$, and the conditions $\tau \in G^+$, $\sigma\tau\sigma^{-1} \in G^+$ are equivalent to each other). It is therefore sufficient to prove III.1.10 in the case where $Z = N$. A representative spinor for Z is then the product of the elements of a base of N, and is homogeneous of degree r. A representative spinor for Z' is of the form $u' = c(\exp v)x_1 \cdots x_h$, with $v \in C_2^N$, $x_1, \cdots, x_h$ forming a base of $Z' \cap N$. Thus, u' is even or odd according as to whether h is even or odd, which proves III.1.10.

III.1.11. *Any totally singular subspace U of dimension $r - 1$ of M is contained in exactly one even and exactly one odd maximal totally singular subspace of M.*

We may transform U into a subspace of N by an operation of G. It will therefore be sufficient to prove III.1.11 when $U \subset N$. Let $(x_1, \cdots, x_{r-1})$ be a base of U. Let Z be a maximal totally singular space containing U. If Z is of the same kind as N, then $\dim(Z \cap N) \equiv r \pmod 2$ and $\dim(Z \cap N) \geq r - 1$, whence $Z = N$. If not, we see in the same way that $Z \cap N = U$. Let then u be a representative spinor for Z; then we have $u = ax_1 \cdots x_{r-1} + bx_1 \cdots x_r$, where x_r is an element of N not in U. Since u is even or odd, $u = ax_1 \cdots x_{r-1}$, and Z is uniquely determined. Conversely, $x_1 \cdots x_{r-1}$ is a pure spinor and represents a maximal totally singular space not of the same kind as N and containing U.

III.1.12. *Let u, u' be pure spinors which are representative for distinct maximal totally singular space Z, Z'. A necessary and sufficient condition for $u + u'$ to be pure is that* $\dim (Z \cap Z') = r - 2$. *If this is so, then the linear combinations $\neq 0$ of u, u' are representative spinors for all maximal totally singular spaces Z'' such that $Z \cap Z'' = Z \cap Z'$ or $Z'' = Z$.*

Since Z may be transformed into N by an operation of G, it is easily seen that it suffices to prove III.1.12 in the case where $Z = N$ and u is the product of the elements of a base of N. Write $u' = c(\exp v)$ $(x_1 \cdots x_h)$, where $v \in C_2^N$ and $(x_1, \cdots, x_h)$ is a base of $Z' \cap N$. If $u + u'$ is pure, it is representative for a space Z'' such that $Z'' \cap N = Z' \cap N$. For, we have $\rho(x) \cdot u = 0$ for all $x \in N$, which shows that the conditions $\rho(x) \cdot u' = 0$, $\rho(x) \cdot (u + u') = 0$ are equivalent to each other. Thus, we have

$$u + u' = c'(\exp v')x_1 \cdots x_h, \qquad c' \in K, \qquad v' \in C_2^N.$$

We have $h < r$ and u is homogeneous of degree r; writing that u', $u + u'$ have the same homogeneous component of degree h, we obtain $c = c'$. The homogeneous components of degree $h + 2$ of u', $u + u'$ are $cvx_1 \cdots x_h$, $cv'x_1 \cdots x_h$. Were $h < r - 2$, then we would have $vx_1 \cdots x_h = v'x_1 \cdots x_h$, from which it would easily follow that $(\exp v)x_1 \cdots x_h = (\exp v')x_1 \cdots x_h$, $u' = u + u'$, which is impossible. Thus, we have $h \geq r - 2$. Since $u + u'$ is even or odd, we have $h \equiv r$ (mod 2) and therefore $h = r - 2$. Conversely, assume that $h = r - 2$. Then we have $u = x_1 \cdots x_{r-2}x_{r-1}x_r$, where x_{r-1}, x_r are suitably selected elements of N, and

$$u + u' = c(\exp v + c^{-1}x_{r-1}x_r)x_1 \cdots x_{r-2};$$

but this clearly equal to $c(\exp (v + c^{-1}x_{r-1}x_r))x_1 \cdots x_{r-2}$, and $u + u'$ is pure. Let Z'' be any maximal totally singular subspace of M such that $Z'' \cap N = Z' \cap N$, and u'' a representative spinor for Z''. Then u'' is a multiple of $x_1 \cdots x_{r-2}$, and its homogeneous component of degree $r - 2$ is of the form $ax_1 \cdots x_{r-2}$, $a \in K$, while its homogeneous component of degree $r - 1$ is 0. Thus, $u'' - ac^{-1}u'$ is homogeneous of degree r and therefore a scalar multiple of u.

3.2. A Bilinear Invariant

Let α be the main antiautomorphism of C. If Z is any subspace of M, α clearly transforms into itself the subalgebra C^Z generated by Z. If

Z is totally singular, then C^Z is isomorphic to the exterior algebra of Z; if $z_1, \cdots, z_h \in Z$, then we have

$$\alpha(z_1 \cdots z_h) = z_h \cdots z_1 = (-1)^{h(h-1)/2} z_1 \cdots z_h .$$

It follows that α multiplies every homogeneous element of degree h of C^Z by $(-1)^{h(h-1)/2}$.

We shall apply this to the case where $Z = N$. If $u, v \in C^N$, then we have

$$\alpha(uf)vf = \alpha(f)\alpha(u)vf = (-1)^{r(r-1)/2} f\alpha(u)vf,$$

since f is homogeneous of degree r in C^P. We have $\alpha(u)v \in C^N$ and $f\alpha(u)vf = (\rho(f)\cdot\alpha(u)v)f$. Let us now determine the operation $\rho(f)$. Let e be the product of the elements of a base $(x_1, \cdots, x_r)$ of N, and let $(y_1, \cdots, y_r)$ be the base of P such that $B(x_i, y_j) = \delta_{ij}$. Since $\rho(y_i)$ is a homogeneous operator of degree -1, $\rho(f)$ is of degree $-r$ and maps upon 0 every homogeneous element of degree $< r$ of C^N. On the other hand, $\rho(y_i)$ maps x_i upon 1 and x_j upon 0 if $i \neq j$; it follows easily, since each $\rho(y_i)$ is an antiderivation, that $\rho(y_1 \cdots y_r)$ maps e upon $(-1)^{r(r-1)/2}\cdot 1$. We have $f = cy_1 \cdots y_r$, c a scalar $\neq 0$. Thus, we see that, if de is the homogeneous component of degree r of $\alpha(u)v$, then

$$\rho(f)\cdot\alpha(u)v = (-1)^{r(r-1)/2} cd\cdot 1.$$

We may obviously select e in such a way that $c = 1$. This being done, *we denote by* $\beta(u, v)e$ *the homogeneous component of degree* r *of* $\alpha(u)v$, whence

$$\alpha(uf)vf = \beta(u, v)f. \tag{1}$$

It is clear that β is a bilinear form on $S \times S$ ($S = C^N$ being the space of spinors).

III.2.1. *Let* λ *be the norm homomorphism of the Clifford group* Γ. *Then we have, for* $s \in \Gamma$, $u, v \in S$,

$$\beta(\rho(s)\cdot u, \rho(s)\cdot v) = \lambda(s)\beta(u, v).$$

For we have $(\rho(s)\cdot u)f = suf$, $(\rho(s)\cdot v)f = svf$, and

$$\alpha(suf)svf = \alpha(uf)\alpha(s)svf = \lambda(s)\alpha(uf)vf.$$

III.2.2. *Let* x *be in* M, u *and* v *in* S. *Then we have*

$$\beta(\rho(x)\cdot u, \rho(x)\cdot v) = Q(x)\beta(u, v),$$

$$\beta(\rho(x)\cdot u, v) = \beta(u, \rho(x)\cdot v).$$

The first formula is obtained by exactly the same computation that was used in proving III.2.1. We have $(\rho(x)\cdot u)f = xuf$, $\alpha(xuf)vf = \alpha(uf)\alpha(x)vf = \alpha(uf)xvf$, which proves the second formula.

It follows from III.2.1 that β is a bilinear invariant of the spin representation of Γ_0^+.

The form β is nondegenerate. For, if u is an element $\neq 0$ in the exterior algebra C^N, we have $\alpha(u) \neq 0$ and there is a $v \in C^N$ such that $\alpha(u)v = e$, whence $\beta(u, v) = 1$.

Since α is an antiautomorphism, α^2 is an automorphism. Since $\alpha^2(x) = x$ for $x \in M$, α^2 is the identity, and $\alpha(\alpha(u)v) = \alpha(v)u$. Since $\alpha(e) = (-1)^{r(r-1)/2}e$, we have

$$\beta(v, u) = (-1)^{r(r-1)/2}\beta(u, v). \tag{2}$$

III.2.3. *Let S_p, S_i be the spaces of even and odd half-spinors. If $r \equiv 0 \pmod 2$, then β is zero on $S_p \times S_i$ and $S_i \times S_p$; if $r \equiv 1 \pmod 2$, then β is zero on $S_p \times S_p$ and on $S_i \times S_i$.*

For, if u, v are homogeneous of degrees δ_u, δ_v, then $\beta(u, v) = 0$ if $\delta_u + \delta_v \neq r$, and, in particular, if $\delta_u + \delta_v$ has not the same parity as r.

III.2.4. *Let Z and Z' be maximal totally singular subspaces of M, and u, u' representative spinors for Z, Z'. A necessary and sufficient condition for $Z \cap Z'$ to be $\neq \{0\}$ is that $\beta(u, u') = 0$.*

There is an operation $\sigma \in G$ such that $\sigma(Z) = N$; let s be in Γ and such that $\chi(s) = \sigma$; $\rho(s)\cdot u$ and $\rho(s)\cdot u'$ are then representative spinors for $\sigma(Z) = N$ and $\sigma(Z')$. By III.2.1, it will be sufficient to prove our assertion in the case where $Z = N$, $u = e$. In that case, we have $e(\exp v) = e$ for any homogeneous v of degree 2, and the result follows immediately from III.1.

Besides α, we may consider the antiautomorphism $\tilde{\alpha}$ product of α by the main involution of C: $\tilde{\alpha}$ transforms any $x \in M$ into $-x$. If u, $v \in M$, denote by $\tilde{\beta}(u, v)e$ the homogeneous component of degree r of $\tilde{\alpha}(u)v$. We have $\tilde{\alpha}(f) = (-1)^{r(r+1)/2}f$, $\tilde{\alpha}(e) = (-1)^{r(r+1)/2}e$, and we see as above that $\tilde{\alpha}(uf)vf = \tilde{\beta}(u, v)f$. If $s \in \Gamma$, then $\tilde{\alpha}(s) = \alpha(s)$ if $s \in \Gamma^+$, $\tilde{\alpha}(s) = -\alpha(s)$ if s is odd. Proceeding as above, we show that

$$\tilde{\beta}(\rho(s)\cdot u, \rho(s)\cdot v) = \epsilon(s)\lambda(s)\tilde{\beta}(u, v) \tag{3}$$

if $s \in \Gamma$, u, $v \in S$, where $\epsilon(s)$ is -1 if s is odd, $+1$ in the opposite case. Moreover, if $x \in M$, then we have

$$\tilde{\beta}(\rho(x)\cdot u, \rho(x)\cdot v) = -Q(x)\tilde{\beta}(u, v), \tag{4}$$

$$\tilde{\beta}(\rho(x)\cdot u, v) = -\tilde{\beta}(u, \rho(x)\cdot v). \tag{5}$$

Since $\tilde{\alpha}(u) = \alpha(u)$ if $u \varepsilon S_p$, $\tilde{\alpha}(u) = -\alpha(u)$ if $u \varepsilon S_i$, we see that $\tilde{\beta}$ coincides with β on $S_p \times S$, with $-\beta$ on $S_i \times S$.

III.2.5. *The only bilinear invariants of the spin representation of* Γ^+_0 *are the linear combinations of* β, $\tilde{\beta}$, *unless* $m = 2$ *and* K *has either* 2 *or* 3 *elements.*

Let β' be a bilinear invariant of the spin representation ρ_0^+ of Γ_0^+. Let S^* be the dual space of S, and ω the representation of Γ_0^+ on S^* contragredient to ρ_0^+ (i.e., if $s \varepsilon \Gamma_0^+$, $\omega(s)$ is the transpose of $\rho_0^+(s^{-1})$). There are associated to β, β' linear mappings φ, φ' of S into S^*, and we have $\omega(s) \circ \varphi = \varphi \circ \rho_0^+(s)$, $\omega(s) \circ \varphi' = \varphi' \circ \rho_0^+(s)$ for $s \varepsilon \Gamma_0^+$; moreover, φ is a linear isomorphism. Thus, $\varphi' = \varphi \circ \psi$, where ψ is an automorphism of S which commutes with every operation of $\rho_0^+(\Gamma_0^+)$. We have seen in the proof of II.4.3 that, barring the exceptional cases of the statement of III.2.5, Γ_0^+ is a set of generators of C_+ . Let ρ_p^+, ρ_i^+ be the representations of C_+ on the spaces S_p , S_i . Then $\rho_p^+(C_+)$, $\rho_i^+(C_+)$ are algebras of dimension 2^{2r-2} (they are isomorphic to the simple ideals of C_+), and S_p , S_i are of dimension 2^{r-1} . Thus, $\rho_p^+(C_+)$ and $\rho_i^+(C_+)$ are the algebras of *all* endomorphisms of S_p and S_i , which shows that the representations of Γ_0^+ on S_p , S_i are absolutely simple. Besides, these representations are inequivalent to each other (II.4.3). It follows immediately that the algebra of endomorphisms ψ of S which commute with all operations of $\rho_0^+(\Gamma_0^+)$ is of dimension 2. (It is spanned by I and by the operator which maps the elements of S_p upon themselves, those of S_i upon 0.) Thus, the space of bilinear invariants of ρ_0^+ is of dimension 2, and is therefore spanned by β, $\tilde{\beta}$, since β, $\tilde{\beta}$ are obviously linearly independent.

III.2.6. *Let* Z *be a maximal totally singular subspace of* M *and* σ *an operation of* G *such that* $\sigma(Z) = Z$. *Let* σ_Z *be the restriction of* σ *to* Z. *Then there exists an* $s \varepsilon \Gamma^+$ *such that* $\chi(s) = \sigma$, $\lambda(s) = \det \sigma_Z$.

Let τ be an operation of G which transforms Z into N. Then $\sigma' = \tau\sigma\tau^{-1}$ transforms N into itself, and, if σ'_N is the restriction of σ' to N, then $\det \sigma'_N = \det \sigma_Z$. Let $t \varepsilon \Gamma$ be such that $\chi(t) = \tau$; if $\chi(s') = \sigma'$, then we have $\chi(t^{-1}s't) = \sigma$, $\lambda(t^{-1}s't) = \lambda(s')$, and $t^{-1}s't \varepsilon \Gamma^+$ if $s' \varepsilon \Gamma^+$. Thus, we see that we may assume that $Z = N$. Let s_1 be any element of Γ such that $\chi(s_1) = \sigma$. Since $\sigma(N) = N$, σ is in G^+ and $s_1 \varepsilon \Gamma^+$. The automorphism $w \rightarrow s_1ws_1^{-1}$ of C transforms N into itself; thus, we have $s_1C^Ns_1^{-1} = C^N$. If $u \varepsilon S = C^N$, we may write $s_1uf = s_1us_1^{-1}s_1f$; we have $s_1f = (\rho(s_1)\cdot 1)f$, whence $\rho(s_1)\cdot u = (s_1us_1^{-1})\,(\rho(s_1)\cdot 1)$. Now, 1 is a representative spinor for P; thus, $\rho(s_1)\cdot 1$ is a representative spinor for $\sigma(P)$.

Since $P \cap N = \{0\}$, $\sigma(P) \cap N = \{0\}$ and $\rho(s_1)\cdot 1 = c\,(\exp v)$, $v \in C_N^{+}$ (by III.1.9). The mapping $u \to s_1 u s_1^{-1}$ is the automorphism of C^N which extends σ_N, whence $s_1 e s_1^{-1} = (\det \sigma_N)e$; moreover, $s_1 1 s_1^{-1} = 1$. Thus, we have $\rho(s_1)\cdot 1 = c \exp v$, $\rho(s_1)\cdot e = c\,(\det \sigma_N)e$. We have $\beta(\rho(s_1)\cdot 1, \rho(s_1)\cdot e) = \lambda(s_1)\,\beta(1, e)$; on the other hand, we have $\exp v \in \Gamma_0^{+}$ (III.1.7), whence $\beta\,(\exp v, (\exp v)e) = \beta(1, e)$. Since $\beta(1, e) \neq 0$, we have $c^2\det \sigma_N = \lambda(s_1)$, and $s = c^{-1}s_1$ has the required property.

III.2.7. *Any two maximal totally singular subspaces Z, Z' of M may be transformed into each other by an operation of $G_0 = \chi(\Gamma_0)$.*

Let σ be an element of G such that $\sigma(Z') = Z$ and s an element of Γ such that $\chi(s) = \sigma$. It will be sufficient to prove that there is an $s' \in \Gamma$ such that $(\chi(s'))\,(Z) = Z$, $\lambda(s') = (\lambda(s))^{-1}$. We can find an automorphism of Z of determinant $(\lambda(s))^{-1}$; this automorphism is a Q-automorphism and may be extended to an operation $\sigma' \in G$. It follows from III.2.6 that there is an $s' \in \Gamma$ such that $\chi(s') = \sigma'$, $\lambda(s') = (\lambda(s))^{-1}$; s' has the required properties.

In the case where K is not of characteristic 2 and $r(r-1) \equiv 0 \pmod 4$, β is symmetric and, if we set $\gamma(u) = \frac{1}{2}\beta(u, u)$, γ is a quadratic form on S whose associated bilinear form is β. It is clear that

$$\gamma(\rho(s)\cdot u) = \lambda(s)u \qquad (s \in \Gamma,\ u \in S), \tag{6}$$

$$\gamma(\rho(x)\cdot u) = Q(x)\gamma(u) \qquad (x \in M,\ u \in S). \tag{7}$$

Moreover, γ is of maximal index. For, let $(x_1, \cdots, x_r)$ be a base of N. For each subset $\{i_1, \cdots, i_h\} = H$ of $\{1, \cdots, r\}$, with $i_1 < \cdots < i_h$, let

$$\xi(H) = x_{i_1} \cdots x_{i_h};$$

these elements form a base of S. Barring the trivial case where $r = 0$, it is easily seen that we can form a set $\{H_k\}$ of 2^{r-1} of the sets H, no two of which are complementary to each other. We then have $\beta(\xi(H_k), \xi(H_l)) = 0$ $(1 \leq k, l \leq 2^{r-1})$, and γ vanishes on the space ζ spanned by the H_k's. Since $\dim \zeta = 2^{r-1}$, γ is of index 2^{r-1}. If $r \equiv 1 \pmod 4$, then γ is zero on S_p, S_i. If $r \equiv 0 \pmod 4$, then the restrictions of γ to S_p, S_i are of rank 2^{r-1} and of index 2^{r-2}, for we see immediately that our set $\{H_k\}$ must contain 2^{r-2} elements of even cardinal numbers and 2^{r-2} elements of odd cardinal numbers.

Assume now that K is of characteristic 2. We shall see that, provided $r > 2$, there is a quadratic form γ on S with properties similar to the one constructed above in the case where K is not of characteristic 2 and $r(r-1) \equiv 0 \pmod 4$. We make use of the operation of "reduced squaring" in C^N, introduced by G. Papy[1]. Using the same notation as above, we further number the elements $\xi(H)$ by indices $0, 1, \cdots, 2^r - 1$ so as to arrange them in a sequence $(\xi_0, \cdots, \xi_p)$ $(p = 2^r - 1)$ such that $\xi_0 = 1, \xi_i = x_i$ for $1 \leq i \leq r$. If

$$u = \sum_{i=0}^{p} a_i \xi_i , \qquad a_i \in K,$$

we set

$$u^{[2]} = \sum_{i<j} a_i a_j \xi_i \xi_j + a_0^2.$$

It is clear that $(au)^{[2]} = a^2 u^{[2]}$ if $a \in K$. Since $\xi_i^2 = 0$ for $i > 0$, an easy computation shows that $(u + v)^{[2]} = u^{[2]} + uv + v^{[2]}$ for $u, v \in S$. From this we deduce by induction on h that

$$(u_1 + \cdots + u_h)^{[2]} = \sum_{k=1}^{h} u_k^{[2]} + \sum_{k<l} u_k u_l \tag{8}$$

for $u_k \in S$, $1 \leq k \leq h$. Let x be in N and

$$u = \sum_{i=0}^{p} a_i \xi_i$$

in S; then we have

$$(xu)^{[2]} = \sum_{i=0}^{p} (x\xi_i)^{[2]},$$

since $(x\xi_i)(x\xi_j) = 0$. Now, write

$$x = \sum_{k=1}^{r} b_k x_k ;$$

each $x_k \xi_i$ is either 0 or a $\xi_{i'}$ of index $i' > 0$: in either case, $(x_k \xi_i)^{[2]} = 0$. On the other hand, we have $(x_k \xi_i)(x_l \xi_i) = 0$ if $i > 0$, since $\xi_i^2 = 0$. Thus, we have

$$(xu)^{[2]} = a_0^2 \sum_{k<l} b_k b_l x_k x_l ,$$

and $(xu)^{[2]} = 0$ if the homogeneous component of degree 0 of u is 0. Now, let $(x'_1, \cdots, x'_r)$ be any other base of M; let ξ'_i $(0 \leq i \leq 2^r - 1)$ be the products

$$x'_{i_1} \cdots x'_{i_h} \qquad (i_1 < \cdots < i_h).$$

[1]G. Papy, "Sur l'arithmétique dans les algèbres de Grassman," Académie Royale de Belgique, Classe des Sciences, *Mémoires*, *26* (1952).

We may assume that $\xi'_0 = 1$, $\xi'_i = x'_i$ for $1 \leq i \leq r$. Then, by what we have just proved, we have $(\xi'_i)^{[2]} = 0$ for $i > r$. If $a'_i \in K$ $(0 \leq i \leq 2^r - 1)$, we find, by (8),

$$\left(\sum_{i=0}^{p} a'_i\xi'_i\right)^{[2]} = a'^2_0 + \sum_{i<j} a'_i a'_j \xi'_i \xi'_j + \sum_{i=1}^{r} a'^2_i (\xi'_i)^{[2]}.$$

This formula shows that, although $u^{[2]}$ may depend on the choice of the base in M, its homogeneous component of degree r does not, provided $r > 2$ (for $(\xi'_i)^{[2]} = (x'_i)^{[2]}$ is homogeneous of degree 2 if $1 \leq i \leq r$). Assuming from now on that $r > 2$, *we denote by* $\gamma(u)e$ *the homogeneous component of degree* r *of* $u^{[2]}$; γ is a quadratic form on S.

Since K is of characteristic 2, α induces the identity automorphism of $C^N = S$, whence $\alpha(u)v = uv$, and $\beta(u, v)e$ is the homogeneous component of the degree r of uv. Since $(u + v)^{[2]} = u^{[2]} + uv + v^{[2]}$, we have

$$\gamma(u + v) = \gamma(u) + \gamma(v) + \beta(u, v), \tag{9}$$

which shows that β is the bilinear form associated to γ.

It follows from the computation made above that $\gamma(xu) = 0$ for any $x \in N$, $u \in S$. We shall see that we have also $\gamma(\rho(y)\cdot u) = 0$ for every $y \in P$. We may obviously assume $y \neq 0$; let $(x_1, \cdots, x_r)$ be a base of N such that $B(x_i, y) = 0$ for $i > 1$, $B(x_1, y) = 1$. Since $\rho(y)$ is an antiderivation which maps any $x \in N$ upon $B(x, y)\cdot 1$, we see that, if $i_1 < \cdots < i_p$, $\rho(y)$ maps $x_{i_1} \cdots x_{i_p}$ upon 0 if $i_1 > 1$, upon $x_{i_2} \cdots x_{i_p}$ if $i_1 = 1$. If we write

$$\rho(y)\cdot u = \sum_{i=0}^{p} b_i\xi_i$$

(in the notation used above), then $b_i = 0$ whenever x_1 is one of the factors of ξ_i, and it follows immediately that the homogeneous component of degree r of $(\rho(y)\cdot u)^{[2]}$ is 0, which proves our assertion.

We shall now prove that formula (7) is true in our case for any $x \in M$. The mapping $u \to \gamma(\rho(x)\cdot u) - Q(x)\,\gamma(u)$ is obviously a quadratic form γ' on S. It follows from (9) and III.2.2 that the associated bilinear form of γ' is 0, i.e., that γ' is quasi-linear. To prove that $\gamma' = 0$, it will therefore be sufficient to show that $\gamma'(u) = 0$ for all elements u of a subset of S which spans S. In particular, it will be sufficient to prove that $\gamma'(\xi_i) = 0$ $(0 \leq i \leq p)$, in the notation introduced above. It is clear that $\gamma(\xi_i) = 0$; we have therefore to prove that $\gamma(\rho(x)\cdot\xi_i) = 0$. This is clear if ξ_i is either 1 or x_i $(1 \leq i \leq r)$. If ξ_i is a product of $h > 1$ factors x_i, we write $x = x_N + x_P$, $x_N \in N$, $x_P \in P$; since $h > 1$, we may write ξ_i in the form $x'u'$, where x' is an element of N such that $\beta(x_P, x') =$

0 and u' a product of $h - 1$ factors in N. Then $\rho(x) \cdot \xi_i$ is $x_N x' u' + x' \cdot \rho(x_P) \xi_i = x'v'$, with $v' \in S$, and we have seen that the homogeneous component of degree r of $(x'v')^{[2]}$ is 0 for all $x' \in N$, $v' \in S$. Thus, (7) is true for our form γ.

Since $r > 2$, $m > 4$, it follows from II.3.1 that Γ is generated by the elements of $\Gamma \cap M$. Since $\lambda(x) = Q(x)$ if $x \in \Gamma \cap M$, it follows immediately from (7) that (6) is true for all $s \in \Gamma$.

The form γ is still of maximal index 2^{r-1}, and, if r is even, the restrictions of γ to S_p and S_i are of maximal index 2^{r-2}. The proofs of these assertions are exactly the same as in the case where K is not of characteristic 2.

If $r = 2$, K of characteristic 2, there is no quadratic form γ on S for which (9) and (7) hold. For, let (x_1, x_2) be a base of N such that $x_1 x_2 = e$. By (7), $\gamma(x) = Q(x)\,\gamma(1) = 0$ for all $x \in N$; thus, $0 = \gamma(x_1 + x_2) = \beta(x_1, x_2)$, while $\beta(x_1, x_2)$ is 1.

Remark. If s is any element of C (not necessarily in Γ) such that $\alpha(s)s$ is a scalar multiple $\lambda \cdot 1$ of 1, then we have

$$\beta(\rho(s)\cdot u,\ \rho(s)\cdot v) = \lambda\beta(u, v) \qquad (u, v \in S);$$

this is proved in the same manner that we proved III.2.1. Similarly, if s is such that $\tilde{\alpha}(s)s = \lambda \cdot 1$, then we have

$$\tilde{\beta}(\rho(s)\cdot u,\ \rho(s)\cdot v) = \lambda\tilde{\beta}(u, v).$$

3.3. The Tensor Product of the Spin Representation with Itself

We consider now the space $S \otimes S$, tensor product of the space S of spinors with itself. This is the space of a representation $\rho \otimes \rho$ of Γ, tensor product of ρ with itself, which is defined by the condition that

$$(\rho \otimes \rho)(s)\cdot u \otimes v = (\rho(s)\cdot u) \otimes (\rho(s)\cdot v)$$

for $s \in \Gamma$, u, v, $\in S$.

If $s \in \Gamma$, then we know that $\alpha(s)s = \lambda(s)\cdot 1$, $\lambda(s)$ being a scalar $\neq 0$.

III.3.1. *The representation $\rho \otimes \rho$ of the group Γ is equivalent to the representation which assigns to every $s \in \Gamma$ the endomorphism $w \to \lambda(s) s w s^{-1}$ of the vector space C.*

The mapping $(u, v) \to u f \alpha(v)$ of $S \times S$ into C is clearly bilinear; as such, it defines a linear mapping φ of the tensor product $S \otimes S$ into C such that $\varphi(u \otimes v) = u f \alpha(v)$ for any $u, v \in S$. We have $\varphi(S \otimes S) = C$. For, let w be in C; then $wuf = (\rho(w)\cdot u)f$, whence $wuf\alpha(v) = (\rho(w)\cdot u) f\alpha(v) \in \varphi(S \otimes S)$. On the other hand, we have $\alpha(f)\alpha(v)\alpha(w) = \alpha(wvf) = \alpha(f)\alpha(\rho(w)\cdot v)$ and $\alpha(f) = \pm f$, whence $\alpha(f)\alpha(v)\alpha(w) \in \varphi(S \otimes S)$. Since

α is a mapping of C onto itself, we conclude that $\varphi(S \otimes S)$ is a two-sided ideal, obviously $\neq \{0\}$, in C. Since C is simple, $\varphi(S \otimes S) = C$. But $S \otimes S$ and C have the same dimension 2^m; φ is therefore an isomorphism. Moreover, if $s \in \Gamma$, then we have $\varphi(\rho(s)\cdot u, \rho(s)\cdot v) = sufa(v)\alpha(s) = \lambda(s)sufa(v)s^{-1}$, which proves III.3.1.

We shall identify the space $S \otimes S$ to C by means of the mapping φ introduced in the proof of III.3.1. For any $h \geq 0$, let C_h be the space spanned by the products of at most h elements of C; it is then clear that C_h is mapped into itself by the operations of $(\rho \otimes \rho)(\Gamma)$. Thus, C_h/C_{h-1} is the space of a representation θ_h of Γ: if $\bar{w} \in C_h/C_{h-1}$ is the coset modulo C_{h-1} of a $w \in C_h$, then $\theta_h(s)\cdot\bar{w}$ is the coset of $(\rho \otimes \rho)(s)\cdot w$. It follows immediately from III.3.1 that $\theta_h(s) = \lambda(s)\theta_h'(\chi(s))$, where θ_h' is a representation of $\chi(\Gamma) = G$ (χ being the vector representation of Γ, whose kernel is the intersection of Γ with the center of C). We shall see that *θ'_h is equivalent to the representation of G on the h-vectors.*

We define a bilinear form B_0 on $M \times M$ such that $B_0(x, x) = Q(x)$ in the manner indicated in the proof of II.2.1. Thus, B_0 is zero on $N \times N$, on $N \times P$, and on $P \times P$ and coincides with B on $P \times N$. Making use of B_0, we identify the underlying vector space of C with that of the exterior algebra E of M in the manner described in II.1.6. Let E_h be the space of homogeneous elements of degree h of E. Then we have $C_h = \sum_{h' \leq h} E_{h'}$, $C_{h-1} = \sum_{h' \leq h-1} E_{h'}$, so that C_h is the direct sum of C_{h-1} and E_h. Let s be in Γ, and $x_1, \cdots, x_h$ in M; set $\sigma = \chi(s)$, $x'_i = \sigma\cdot x_i = xs_is^{-1}$ $(1 \leq i \leq h)$ and denote by ζ_h the representation of G on the h-vectors. Then we have

$$(\rho \otimes \rho)(s)\cdot(x_1 \cdots x_h) = \lambda(s)x'_1 \cdots x'_h,$$

and $\theta'_h(\sigma)$ transforms the coset of $x_1 \cdots x_h$ modulo C_{h-1} into that of $x'_1 \cdots x'_h$. On the other hand, $\zeta_h(\sigma)$ transforms $x_1 \wedge \cdots \wedge x_h$ into $x'_1 \wedge \cdots \wedge x_h'$. Now, we have

$$x_1 \cdots x_h \equiv x_1 \wedge \cdots \wedge x_h \qquad (\text{mod } C_{h-1}),$$

$$x'_1 \cdots x'_h \equiv x'_1 \wedge \cdots \wedge x_h' \qquad (\text{mod } C_{h-1}),$$

and the elements of E_h form a complete system of representatives for the elements of C_h modulo C_{h-1}. It follows immediately that θ'_h is equivalent to ζ_h.

III.3.2. *Let u be an element $\neq 0$ of S. In order for u to be a pure spinor, it is necessary and sufficient that the following conditions be satisfied: (a) u is either even or odd; (b) $u \otimes u = uf\alpha(u)$ belongs to the space C_r spanned by the products of r elements of M. If u is a representative* *spinor of a maximal totally singular space Z, then $uf\alpha(u)$ is the product of the elements of a base of Z.*

Assume first that u is representative of Z. Let s be an element of Γ such that $sPs^{-1} = Z$. Then we have $uf = asf$, $a \in K$, $a \neq 0$, and $\alpha(f)\,\alpha(u) = a\alpha(f)\alpha(s) = a\lambda(s)\alpha(f)s^{-1}$; but $\alpha(f) = \pm f$, whence $f\alpha(u) = a\lambda(s)fs^{-1}$ and $uf\alpha(u) = a^2\lambda(s)sfs^{-1}$. It is clear that sfs^{-1} is the product of the elements of a base of Z and therefore belongs to C_r.

Now, let Σ be the set of elements $u \neq 0$ of S which satisfy conditions (*a*) and (*b*). Then Σ contains the set of pure spinors. Let u be in Σ and s in Γ; then $\rho(s)\cdot u$ is either even or odd and

$$(\rho(s)\cdot u)f\alpha(\rho(s)\cdot u) = \lambda(s)s(uf\alpha(u))s^{-1}$$

by the proof of III.3.1, whence $\rho(s)\cdot u \in \Sigma$. Assume first that we are not considering the exceptional case where $r = 1$, K has only 2 elements, and Q is of index 1. Then ρ is simple (II.4.1). It follows that S is spanned by the elements of the form $\rho(s)\cdot u$. Now, S is the exterior algebra E^N of the space N; since the elements $\rho(s)u$ span Σ, we see that one of them has a homogeneous component of degree 0 in E^N which is $\neq 0$. In order to prove that u is a pure spinor, it suffices to prove that $\rho(s)\cdot u$ is one; thus, we may assume that the homogeneous component of degree 0 of u is $\neq 0$, and even (by multiplication by a scalar $\neq 0$) that this component is 1. Since u is either even or odd, u is then even. Let u_h be its homogeneous component of degree h; then we have $u_h = 0$ for h odd. The homogeneous component of degree 2 of the element $(\exp u_2)^{-1}$ of Γ is $-u_2$; that of $(\exp u_2)^{-1}u$ is therefore 0. Since $(\exp u_2)^{-1} \in \Gamma$, we see that we may restrict ourselves without loss of generality to the case where $u_2 = 0$. We shall then prove that $u = 1$, which will establish that u is a pure spinor.

To do this, we first prove that, if $x \in N$, $u \in \Sigma$, $xu \neq 0$, then $xu \in \Sigma$. It is clear that xu is either even or odd and that $xuf\alpha(xu) = x(uf\alpha(u))x$. Thus, we have only to prove that $xC_rx \in C_r$. Let $z_1, \cdots, z_r$ be in M; then we have, by II.1.6

$$xz_1 \cdots z_rx \equiv x \wedge z_1 \wedge \cdots \wedge z_r \wedge x = 0 \qquad (\text{mod } C_r),$$

which proves our assertion.

Return now to the case where $u_0 = 1$, $u_2 = 0$. Were $u \neq 1$, then there would exist a smallest $h > 0$ such that $u_h \neq 0$, and h would be ≥ 4. Let then w be a decomposable element of S, homogeneous of degree $r - h$, such that $wu_h = e$. (It will be remembered that e is the product of the elements of a certain base of N.) Thus, $wu = w + e$ would be in

Σ. The proof will therefore be complete if we show that, w being a homogeneous decomposable element $\neq 0$ of degree $h \leq r - 4$, $w + e$ cannot be in Σ. There exist bases $(x_1, \cdots, x_r)$ of N, $(y_1, \cdots, y_r)$ of P such that

$$w = x_1 \cdots x_h \qquad e = x_1 \cdots x_r \qquad f = y_1 \cdots y_r \qquad B(x_i, y_j) = \delta_{ij} .$$

We have $(w + e)f(\alpha(w + e)) = wf\alpha(w) + ef\alpha(e) + wf\alpha(e) + ef\alpha(w)$. We know that w and e are pure spinors (III.1.9); it will therefore be sufficient to prove that $wf\alpha(e) + ef\alpha(w)$ is not in C_r , Let N_0 be the space spanned by $x_1, \cdots, x_h$; by III.1.8, $x_1 \cdots x_h = w$ is a representative spinor for $N_0 + P_0$, where P_0 is the space of elements of P which are orthogonal to those of N_0 . This space is spanned by $y_{h+1}, \cdots, y_r$. It follows from the part of III.3.2 which has been proved already that $wf\alpha(w) = ax_1 \cdots x_h y_{h+1} \cdots y_r$, $a \neq 0$ in K. Let $w' = x_{h+1} \cdots x_r$; then we have $e = (-1)^{h(r-h)} w'w$, $\alpha(e) = (-1)^{h(r-h)} \alpha(w)\alpha(w')$, and

$$wf\alpha(e) + ef\alpha(w)$$
$$= ax_1 \cdots x_h[(-1)^{h(r-h)} y_{h+1} \cdots y_r x_r \cdots x_{h+1} + x_{h+1} \cdots x_r y_{h+1} \cdots y_r].$$

For any $z \in M$, let $L(z)$ be the operator of left multiplication by z in E and $\delta(z)$ the antiderivation of E which maps any $z' \in M$ upon $B_0(z, z')\cdot 1$. Then the operator of multiplication by z in C is $L(z) + \delta(z)$. We have

$$y_{h+1} \cdots y_r x_r \cdots x_{h+1} = \prod_{i=h+1}^{r} (L(y_i) + \delta(y_i))\cdot x_r \cdots x_{h+1} .$$

If $i \neq j$, $v \in E$, then we have $\delta(y_i)y_j = 0$, whence $\delta(y_i)\cdot y_j \wedge v = - y_j \wedge \delta(y_i)\cdot v$, and $\delta(y_i)$ anticommutes with $L(y_j)$. Thus, we have

$$\prod_{i=h+1}^{r} (L(y_i) + \delta(y_i)) = \prod_{i=h+1}^{r} L(y_i) + \sum_{i=h+1}^{r} (-1)^{r-i} P_i \delta(y_i) + \Lambda,$$

where P_i is the product deduced from $L(y_{h+1}) \cdots L(y_r)$ by omitting the factor $L(y_i)$ and where Λ is a sum of homogeneous operators whose degrees are $< r - h - 2$. We have $x_r \cdots x_{h+1} = x_r \wedge \cdots \wedge x_{h+1}$ and $\delta(y_i)\cdot x_r \cdots x_{h+1} = (-1)^{r-i}\xi_i$, where ξ_i is the product deduced from $x_r \wedge \cdots \wedge x_{h+1}$ by omitting the factor x_i . If $x \in N$, we have $\delta(x) = 0$, so that the operators of left multiplication by x in E and C are identical to each other. It follows that the homogeneous component of degree $r + (r - h) - 2$ of $wf\alpha(e) + ef\alpha(w)$ is

$$(-1)^{h(r-h)} ax_1 \wedge \cdots \wedge x_h \cdot \sum_{i=h+1}^{r} P_i \cdot \xi_i$$

and $P_i \cdot \xi_i = \pm \xi_i \wedge \eta_i$, where η_i is the product deduced from $y_{h+1} \wedge \cdots \wedge y_r$ by omitting the factor y_i. This shows that the homogeneous component of degree $r + (r - h - 2)$ of $wf\alpha(e) + ef\alpha(w)$ in E is $\neq 0$. Since $r - h \geq 4$, $wf\alpha(e) + ef\alpha(w)$ is not in C_r, which concludes the proof of III.3.2.

If we consider the exceptional case mentioned above, then S is of dimension 2, S_p and S_i are of dimension 1, and the conclusion that u is pure follows from the assumption that u is even or odd.

III.3.3. *Let Z and Z' be maximal totally singular subspaces of M, u and u' representative spinors for Z and Z', and $h = \dim Z \cap Z'$. Then $uf\alpha(u')$ is in C_{m-h} but not in C_{m-h-1}.*

We first establish

Lemma 1. *Let the notation be as in* III.3.3, *and let Z_1, Z'_1 be maximal totally singular subspaces of M such that* $\dim (Z_1 \cap Z'_1) = h$. *Then there exists an operation $\sigma \in G$ such that $\sigma(Z) = Z_1$, $\sigma(Z') = Z'_1$.*

There is a vector-space isomorphism of Z with Z_1 which transforms $Z \cap Z'$ into $Z_1 \cap Z'_1$; since Z is totally singular, this isomorphism is a Q-isomorphism and may be extended to an operation σ_1 of G. It follows immediately that it is sufficient to consider the case where $Z = Z_1$, $Z \cap Z' = Z \cap Z'_1$. In that case, Lemma 1 follows from III.1.7.

This being said, we can now prove III.3.3. Let $(x_1, \cdots, x_r)$ be a base of N and $(y_1, \cdots, y_r)$ a base of P such that $B(x_i, y_j) = \delta_{ij}$ $(1 \leq i, j \leq r)$, $y_1 \cdots y_r = f$, and let Z'_0 be the maximal totally singular space whose representative spinor is $x_1 \cdots x_h$. Then $Z'_0 \cap N$ is of dimension h, and there exists a $\sigma \in G$ such that $\sigma(Z) = N$, $\sigma(Z') = Z'_0$. Since e is a representative spinor for N, we have $\rho(s) \cdot u = ae$, $\rho(s) \cdot u' = bx_1 \cdots x_h$, where a, b are scalars $\neq 0$; thus, we have

$$abef\alpha(x_1 \cdots x_h) = \lambda(s)s(uf\alpha(u'))s^{-1}.$$

Since the mapping $w \to sws^{-1}$ maps each C_k onto itself, it is sufficient to prove that $ef\alpha(x_1 \cdots x_h) = efx_h \cdots x_1$ is in C_{m-h} but not in C_{m-h-1}. This element may be written as $\pm x_{h+1} \cdots x_r wf\alpha(w)$, where $w = x_1 \cdots x_h$, and it follows from III.3.2 that $wf\alpha(w)$ is the product of the elements of some base of Z_0. Now, Z_0 is spanned by $x_1, \cdots, x_h$ and by those elements of P which are orthogonal to $x_1, \cdots, x_h$ (see III.1.8); thus, $(x_1, \cdots, x_h, y_{h+1}, \cdots, y_r)$ is a base of Z_0. Since Z_0 is totally singular, the algebra generated by it in C is isomorphic to the exterior algebra of Z_0, and the products of the elements of the various bases of Z_0 differ only from each other by scalar factors $\neq 0$. This shows that

$$(x_1 \cdots x_h)f\alpha(x_1 \cdots x_h) = cx_1 \cdots x_h y_{h+1} \cdots y_r , \tag{1}$$

where c is a scalar $\neq 0$. Thus, we have

$$ef\alpha(x_1 \cdots x_h) = c'ey_{h+1} \cdots y_r , \tag{2}$$

where c' is a scalar $\neq 0$. This shows that $ef\alpha(x_1 \cdots x_h)$ is in C_{m-h} and is congruent mod C_{m-h-1} to the element $c'x_1 \wedge \cdots \wedge x_r \wedge y_{h+1} \wedge \cdots \wedge y_r$, which is an element $\neq 0$ in E_{m-h}; III.3.3 is thereby proved.

3.4. The Tensor Product of the Spin Representation with Itself (Characteristic $\neq 2$)

We shall assume in this section that K is not of characteristic 2.

Since K is of characteristic $\neq 2$, we may make use of the bilinear form $B_0 = \frac{1}{2}B$ on $M \times M$, which has the property that $B_0(x, x) = Q(x)$. Making use of II.1.6, we shall identify C with the exterior algebra E of M by making use of the bilinear form $B/2$. This identification is different from the one used in Section 3. But now the algebra E may be defined in terms of C and M alone, without making use of a special choice of totally singular subspaces N and P. As a consequence, any automorphism j of C with transforms M into itself will also be an automorphism of E. Let us prove this point more explicitly.

If $x \in M$, denote by δ_x the antiderivation of E which maps any $y \in M$ upon $\frac{1}{2}B(x, y)\cdot 1$, and by L_x, L'_x the operators of left multiplication by x in E and C, whence $L'_x = L_x + \delta_x$. Let x and y be in M. Since δ_x is an antiderivation of E and y homogeneous of degree 1, we have $\delta_x L_y + L_y \delta_x = \frac{1}{2}B(x, y)\, I$, where I is the identity. On the other hand, we know that $\delta_z^2 = 0$ for every $z \in M$; applying this to x, y, and $x + y$, we obtain $\delta_x\delta_y + \delta_y\delta_x = 0$. It follows that $\delta_x L'_y + L'_y\delta_x = \frac{1}{2}B(x, y)I$. The operator δ_x is uniquely characterized by the following properties: it is linear, it maps 1 upon 0, and, for any $y \in M$, we have $\delta_x L'_y + L'_y\delta_x = \frac{1}{2}B(x, y)I$. For, let δ be any operator with these properties and $\delta' = \delta_x - \delta$. Let $\mathfrak{a}$ be the set of $u \in C$ such that $\delta'\cdot u = 0$. Then $\mathfrak{a}$ is a vector space containing 1 and M; if $u \in \mathfrak{a}$, then $\delta'\cdot yu = \delta' L'_y u = - L'_y\delta'\cdot u = 0$; since M generates C, it follows immediately that $\mathfrak{a} = C$, $\delta' = 0$. Now, let j be an automorphism of C such that $j(M) = M$. Since j is an automorphism, we have $jL'_y j^{-1} = L'_{j\cdot y}$ for any $y \in M$. Thus, we have

$$j\delta_x j^{-1}\cdot L'_{j\cdot y} + L'_{j\cdot y}\, j\delta_x j^{-1} = \tfrac{1}{2}B(x, y)I.$$

On the other hand, since $xy + yx = B(x, y)\cdot 1$, we have $B(j\cdot x, j\cdot y) = B(x, y)$; we conclude that $j\delta_x j^{-1} = \delta_{j\cdot x}$. Thus, we have

$$jL_x j^{-1} = j(L'_x - \delta_x)j^{-1} = L'_{j\cdot x} - \delta_{j\cdot x} = L_{j\cdot x} .$$

Since M generates E, it follows immediately from this formula that j is an automorphism of E.

We apply this to the case where $j(u) = sus^{-1}$, s being an element of the Clifford group Γ. Since j is an automorphism of E and maps M onto itself, j maps the space E_h of homogeneous elements of degree h of E onto itself for any h. In Section 3.3, we have denoted by θ the representation of G which is defined by the formula

$$\theta(\chi(s))\cdot u = sus^{-1} \qquad \text{for} \qquad s \in \Gamma,$$

and by $\theta_h(\sigma)$ the restriction of θ to C_h (the space spanned by the products of at most h elements of M). We have seen that the representation of G on C_h/C_{h-1} defined in a natural manner by θ_h is equivalent to the representation ζ_h of G on the h-vectors. Since C_h is the direct sum of C_{h-1} and E_h, we see that θ_h is equivalent to the direct sum of θ_{h-1} and ζ_h. Thus, θ is equivalent to the direct sum of $\zeta_0, \zeta_1, \cdots, \zeta_m$.

Let u and v be any elements of S. Then $uf\alpha(v)$ is an element of $C = E$. We set

$$uf\alpha(v) = \sum_{h=0}^{m} \beta_h(u, v),$$

where β_h is the homogeneous component of degree h of $uf\alpha(v)$. Each β_h is then obviously a bilinear mapping of $S \times S$ into E_h, and we have, for $s \in \Gamma$,

$$\beta_h(\rho(s)\cdot u, \rho(s)\cdot v) = \lambda(s)\zeta_h(\chi(s))\cdot\beta_h(u, v), \tag{1}$$

where χ is the vector representation of Γ, ρ its spin representation, and ζ_h the representation of G on h-vectors.

We shall now study the symmetry properties of the mappings β_h. In order to do this, we need the following result:

III.4.1. *The antiautomorphism α of C is also an antiautomorphism of E; it multiplies the elements of E_h by $(-1)^{h(h-1)/2}$.*

Let $x_1, \cdots, x_h$ be mutually orthogonal vectors in M. Then we have

$$x_1 \cdots x_h = x_1 \wedge \cdots \wedge x_h .$$

We prove this by induction on h. It is obvious if $h = 0$ or 1. Assume that $h > 1$ and that our assertion is true for $h - 1$. Then we have

$$x_1(x_2 \cdots x_h) = x_1 \wedge (x_2 \cdots x_h) + \delta(x_2 \cdots x_h),$$

where δ is the antiderivation of E such that $\delta\cdot x = \frac{1}{2}B(x_1, x)$ if $x \in M$. Since $x_1, \cdots, x_h$ are mutually orthogonal, we have $\delta\cdot x_i = 0$ for $i > 1$, whence $\delta \cdot x_2 \cdots x_h = \delta \cdot (x_2 \wedge \cdots \wedge x_h) = 0$, and our formula is true for h. Now, we have $\alpha(x_1 \cdots x_h) = x_h \cdots x_1$, whence

$$\alpha(x_1 \wedge \cdots \wedge x_h) = x_h \wedge \cdots \wedge x_1 = (-1)^{h(h-1)/2} x_1 \wedge \cdots \wedge x_h .$$

Since M has a base composed of mutually orthogonal vectors, E_h is spanned by the elements of the form $x_1 \wedge \cdots \wedge x_h, x_1, \cdots, x_h$ mutually orthogonal. It follows that α multiplies every element of E^h by $(-1)^{h(h-1)/2}$, from which it follows easily that it is an antiautomorphism of E.

This being said, let u, v be in S. Then we have

$$\alpha(uf\alpha(v)) = v\alpha(f)\alpha(u)$$

and $\alpha(f) = (-1)^{r(r-1)/2} f$; thus, we have

$$\beta_h(v, u) = (-1)^{r(r-1)/2} \alpha(\beta_h(u, v))$$

and therefore

$$\beta_h(v, u) = (-1)^{r(r-1)/2 + h(h-1)/2} \beta_h(u, v). \tag{2}$$

The space S is the direct sum of the spaces S_p of even half-spinors and S_i of odd half-spinors. We propose to study $\beta_h(u, v)$ when u, v are half-spinors. If r is even, then $f \in C_+$ and $uf\alpha(v)$ is in C_+ if u, v are of the same kind, in C_- if they are of opposite kinds; if r is odd, then $f \in C_-$ and $uf\alpha(v)$ is in C_- if u, v are of the same kind, in C_+ if they are of opposite kinds. Since $C_+ = \sum_{h \text{ even}} E_h$, $C_- = \sum_{h \text{ odd}} E_h$, we have proved the following statement:

III.4.2. *If $h \equiv r \pmod 2$, then β_h vanishes on $S_p \times S_i$ and on $S_i \times S_p$; if $h \equiv r + 1 \pmod 2$, then β_h vanishes on $S_p \times S_p$ and $S_i \times S_i$.*

Let $(x_1, \cdots, x_r)$ and $(y_1, \cdots, y_r)$ be bases of N and P such that $B(x_i, y_j) = \delta_{ij}$ $(1 \leq i, j \leq r)$, $y_1 \cdots y_r = f$. Let $x'_{2k-1} = x_k - y_k$, $x'_{2k} = x_k + y_k$ $(1 \leq k \leq r)$; then $(x'_1, \cdots, x'_m)$ is a base of M composed of mutually orthogonal vectors, and $Q(x'_i) = (-1)^i$ $(1 \leq i \leq m)$. Let $z = x'_1 \cdots x'_m = x'_1 \wedge \cdots \wedge x'_m$; then z anticommutes with x'_i $(1 \leq i \leq m)$, which shows that z anticommutes with every element of C_- and is in the center of C_+. Let $i_1, \cdots, i_h$ be integers such that $i_1 < \cdots < i_h$; then we have

$$x'_{i_1} \cdots x'_{i_h} z = (-1)^h x'_{j_1} \cdots x'_{j_{m-h}} ,$$

where $\{j_1, \cdots, j_{m-h}\}$ is the complementary set of $\{i_1, \cdots, i_h\}$ and $j_1 < \cdots < j_{m-h}$; in particular, $z^2 = 1$. Comparing with what was said in the proof of I.6.3, we see that the operation of right multiplication by z may also be defined as follows: for any $x \in M$, let $\delta(x)$ be the antiderivation of E which maps any $y \in M$ upon $\frac{1}{2}B(x, y)\cdot 1$; if $\xi_1, \cdots, \xi_h$ are in M, then

$$(\xi_1 \wedge \cdots \wedge \xi_h)z = \delta(\xi_1) \cdots \delta(\xi_h)\cdot z.$$

We have $x'_{2k-1} \wedge x'_{2k} = 2x_k \wedge y_k$, $\delta(y_k)\cdot x_k = \frac{1}{2}\cdot 1$, $\delta(y_k)\cdot y_k = 0$, whence $\delta(y_k)\cdot(x'_{2k-1} \wedge x'_{2k}) = y_k$. On the other hand, we have $\delta(y_k)\cdot x_l = \delta(y_k)\cdot y_l = 0$ if $k \neq l$. It follows easily that

$$fz = \delta(y_1) \cdots \delta(y_r)\cdot z = f.$$

Since $zfz^{-1} = (-1)^r f$, we have $zf = (-1)^r f$.

III.4.3. *Let u be a spinor and v a half-spinor; set $\epsilon = +1$ if v is even, $\epsilon = -1$ if v is odd. Then we have*

$$\beta_{m-h}(u, v) = \epsilon\beta_h(u, v)z.$$

We have

$$ufa(v) = \sum_{h=0}^{m} \beta_h(u, v),$$

whence

$$ufa(v)z = \sum_{h=0}^{m} \beta_h(u, v)z.$$

On the other hand, $\alpha(v)$ is in C_+ if v is even, in C_- if v is odd, whence

$$\begin{aligned} uf\alpha(v)z &= \epsilon ufz\alpha(v) \\ &= \epsilon uf\alpha(v). \end{aligned}$$

Our assertion then follows from the fact that the operation of right multiplication by z transforms E_h into E_{m-h}.

We may now make the results of III.3.2, III.3.3 more precise.

III.4.4. *Let u be an element $\neq 0$ of S. In order for u to be a pure spinor, it is necessary and sufficient that $\beta_k(u, u) = 0$ for all $k \neq r$. Let u and u' be representative spinors for maximal totally singular spaces Z and Z', and let $h = \dim (Z \cap Z')$. Then we have $\beta_k(u, u') = 0$ if $k < h$ or $k > m - h$; $\beta_h(u, u')$ is the exterior product of the elements of some base of $Z \cap Z'$, while $\beta_{m-h}(u, u')$ is the exterior product of the elements of some base of $Z + Z'$.*

If u is a representative spinor of a maximal totally singular space Z, then $uf\alpha(u)$ is the product in C of the elements of a base of Z (by III.3.2);

these elements being mutually orthogonal, their product in C is also their product in E, and, since Z is of dimension r, it belongs to E_r. Assume conversely that $\beta_k(u) = 0$ for $k \neq r$. In order to prove that u is a pure spinor, it suffices in virtue of III.3.2 to show that u is either even or odd. Write $u = u_p + u_i$, where $u_p \in S_p$, $u_i \in S_i$, whence $\beta_k(u, u) = \beta_k(u_p, u_p) + \beta_k(u_i, u_i) + \beta_k(u_p, u_i) + \beta_k(u_i, u_p)$. If $k \not\equiv r \pmod 2$, then we have $\beta_k(u_p, u_p) = \beta_k(u_i, u_i) = 0$, whence $\beta_k(u_p, u_i) + \beta_k(u_i, u_p) = 0$; if $k \equiv r \pmod 2$, then we have $\beta_k(u_p, u_i) = \beta_k(u_i, u_p) = 0$. Thus, we have $\beta_k(u_p, u_i) + \beta_k(u_i, u_p) = 0$ for every k, and

$$u_p \otimes u_i + u_i \otimes u_p = u_p f\alpha(u_i) + u_i f\alpha(u_p)$$

$$= \sum_{k=0}^{m} (\beta_k(u_p, u_i) + \beta_k(u_i, u_p)) = 0.$$

Were u_p and $u_i \neq 0$, then they would be linearly independent and $u_p \otimes u_i + u_i \otimes u_p$ could not be 0. Thus, u is either even or odd.

We proceed now to prove the second assertion of III.4.4. Let $(x_1, \cdots, x_r)$ and $(y_1, \cdots, y_r)$ be bases of N and P such that $B(x_i, y_j) = \delta_{ij}$ $(1 \leq i, j \leq r)$, $y_1 \cdots y_r = f$; set $x_1 \cdots x_r = e$. The space Z_0 spanned by $x_1, \cdots, x_h, y_{h+1}, \cdots, y_r$ is totally singular and $\dim (N \cap Z_0) = h$. It results from Lemma 1, III.3, that there is a $\sigma \in G$ such that $\sigma(Z) = N$, $\sigma(Z') = Z_0$; let s be an operation of Γ such that $\sigma = \chi(s)$. We know that e and $x_1 \cdots x_h$ are representative spinors for N and Z_0 (see III.1.9). Thus, $\rho(s)\cdot u$ and $\rho(s)\cdot u'$ are scalar multiples $\neq 0$ of e and $x_1 \cdots x_h$, respectively. We have

$$\beta_k(\rho(s)\cdot u, \rho(s)\cdot u') = \lambda(s)s\beta_k(u, u')s^{-1}.$$

The mapping $w \to sws^{-1}$ is an automorphism of E and transforms a base of $Z \cap Z'$ (respectively: $Z + Z'$) into a base of $N \cap Z_0$ (respectively: $N + Z_0$). Thus, we see that it is sufficient to prove the second assertion of III.4.4 in the case where $u = e$, $u' = x_1 \cdots x_h$. Making use of formula (2), III.3, we then have

$$uf\alpha(u') = cx_1 \cdots x_r y_{h+1} \cdots y_r$$

(where c is a scalar $\neq 0$). This element is in C_{m-h} and is congruent modulo C_{m-h-1} to the exterior product $cx_1 \wedge \cdots \wedge x_r \wedge y_{h+1} \wedge \cdots \wedge y_r$. It follows that $\beta_k(u, u') = 0$ if $k > m - h$, while

$$\beta_{m-h}(u, u') = cx_1 \wedge \cdots \wedge x_r \wedge y_{h+1} \wedge \cdots \wedge y_r;$$

this is the product in E of the elements of a base of $N + Z_0$. Making use of III.4.3, we have $\beta_k(u, u') = 0$ if $k < h$, and

$$\beta_h(u, u') = \pm c(x_1 \wedge \cdots \wedge x_r \wedge y_{h+1} \wedge \cdots \wedge y_r)z.$$

Making use of what was said above about the operator of right multiplication by z, this is

$$\pm c\delta(x_1) \cdots \delta(x_r)\delta(y_{h+1}) \cdots \delta(y_r)\cdot z.$$

But z is a basic element of E_m, and is therefore a scalar multiple $\neq 0$ of $x_1 \wedge \cdots \wedge x_r \wedge y_1 \wedge \cdots \wedge y_r$. We have $\delta(x_i)\cdot x_j = \delta(y_i)\cdot y_j = 0$, $\delta(x_i)\cdot y_j = \frac{1}{2}\cdot 1$ $(1 \leq i, j \leq r)$. Since each $\delta(x)$ is an antiderivation, it follows immediately that

$$\delta(x_1) \cdots \delta(x_r)\delta(y_{h+1}) \cdots \delta(y_r)\cdot z = c'x_1 \wedge \cdots \wedge x_h,$$

where c' is a scalar $\neq 0$. This is the product in E of the elements of a base of $N \cap Z_0$, which completes the proof of III.4.4.

If $\sigma \in G$, let s be an element of Γ such that $\chi(s) = \sigma$, and $\zeta(\sigma)$ the mapping $w \to sws^{-1}$ $(w \in C = S \otimes S)$. Let ζ^+ be the representation of G^+ induced by ζ; if $0 \leq h \leq m$, let ζ_h^+ be the representation of G^+ on the h-vectors. We know that ζ_h^+ is simple if $h \neq r$, while ζ_r^+ is equivalent to the sum of two simple representations $\zeta_r'^+$ and $\zeta_r''^+$ (see I.6.2 and I.6.4). The representation ζ^+ is the sum of the representation ζ_h^+ for $0 \leq h \leq m$; ζ_h^+ is equivalent to ζ_{m-h}^+, so that we may write

$$\zeta^+ \cong 2 \sum_{h=0}^{r-1} \zeta_h^+ + \zeta_r'^+ + \zeta_r''^+.$$

On the other hand, $C = S \otimes S$ is the direct sum of the four spaces $S_p \otimes S_p$, $S_i \otimes S_i$, $S_p \otimes S_i$, $S_i \otimes S_p$, each of which is clearly mapped into itself by the operations of $\zeta^+(G^+)$ (because

$$(\rho(s)\cdot u)f\alpha(\rho(s)\cdot v) = \lambda(s)s(uf\alpha(v))s^{-1}$$

if $s \in \Gamma$, and, if $s \in \Gamma^+$, then $\rho(s)$ maps S_p and S_i into themselves). If $\sigma \in G^+$, let $\zeta_{pp}(\sigma)$, $\zeta_{ii}(\sigma)$, $\zeta_{pi}(\sigma)$, $\zeta_{ip}(\sigma)$, be the restrictions of $\zeta(\sigma)$ to these four spaces. We wish to analyze the representations ζ_{pp}, ζ_{ii}, ζ_{pi}, ζ_{ip} into their simple components. Let E'_r and E''_r be the spaces of ζ'^+_r and ζ''^+_r. We write

$$C = \sum_{h=0}^{r-1} (E_h + E_{m-h}) + E'_r + E''_r.$$

We know that the representations ζ_h^+ $(h \leq r - 1)$, $\zeta_r'^+$, $\zeta_r''^+$ are all inequivalent except if K has only 3 elements and $r = 1$ (I.6.2). Let us leave this trivial exceptional case aside. Then any subspace of C which is mapped into itself by the operations of $\zeta^+(G^+)$ is the sum of its intersections with the spaces $E_h + E_{m-h}$, E'_r, E''_r. On the other hand, it follows from III.4.2 that

$$S_p \otimes S_p + S_i \otimes S_i = \sum_{h \equiv r \pmod 2} E_h ,$$

while

$$S_p \otimes S_i + S_i \otimes S_p = \sum_{h \not\equiv r \pmod 2} E_h .$$

Let h be $\equiv r \pmod 2$ and ξ an element $\neq 0$ of E_h ; write $\xi = \xi' + \xi''$, where $\xi' \in S_p \otimes S_p$, $\xi'' \in S_i \otimes S_i$. If $u'_p \in S_p$, $u \in S$, then we have

$$uf\alpha(u')z = ufz\alpha(u') = uf\alpha(u'),$$

while, if $u'_i \in S_i$, then we have

$$uf\alpha(u')z = -uf\alpha(u').$$

It follows that

$$\xi z = \xi' - \xi''.$$

If $h \neq r$, then ξz, which is $\neq 0$ in E_{m-h} , is linearly independent of ξ and neither ξ' nor ξ'' can be 0. Since ξ' and ξ'' are linear combinations of ξ, ξz, we see that both $S_p \otimes S_p$ and $S_i \otimes S_i$ meet $E_h + E_{m-h}$, which proves that ζ_h^+ occurs in both ζ_{pp} and ζ_{ii} . Since ζ_h^+ occurs exactly twice in $\zeta_{pp} + \zeta_{ii}$, it occurs exactly once in ζ_{pp} and in ζ_{ii} . Consider now the case where $h = r$. Since $z^2 = 1$, the mapping $\xi \to \xi z$ is an automorphism of order 2 of E_r , and E_r is the direct sum of the space $E_{r,p}$ of those ξ's such that $\xi z = \xi$ and of the space $E_{r,i}$ of those $\xi \in E_r$ such that $\xi z = -\xi$. Both these spaces are mapped into themselves by the operations of $\zeta_r(G^+)$ because z commutes with every element of Γ^+. It is clear that $E_{r,p} \subset S_p \otimes S_p$, $E_{r,i} \subset S_i \otimes S_i$. None of the spaces $E_{r,p}$, $E_{r,i}$, can be the whole of E_r . This follows easily from the description given above of the operation of right multiplication by z, but it can also be proved as follows. Were for instance $E_{r,p} = E_r$, then ζ_{pp} would be equivalent to

$$\sum_{h=0}^{r-1}{}_{h \equiv r(2)} \zeta_h^+ + \zeta_r^+$$

and ζ_{ii} to

$$\sum_{h=0}^{r-1}{}_{h \equiv r(2)} \zeta_h^+ ;$$

but this is impossible, since ζ_{pp} and ζ_{ii} clearly have the same degree. It follows that $E_{r,p}$, $E_{r,i}$, are the spaces of the two simple representations $\zeta_r'^+$, $\zeta_r''^+$; from now on we shall denote by ζ_{rp}^+ (respectively: ζ_{ri}^+) the one of the two representations $\zeta_r'^+, \zeta_r''^+$ whose space is $E_{r,p}$ (respectively: $E_{r,i}$). We then obtain the following formulas:

$$\zeta_{pp} \cong \sum_{h=0,\ h\equiv r \pmod 2}^{r-1} \zeta_h^+ + \zeta_{r,p}^+,$$

$$\zeta_{ii} \cong \sum_{h=0,\ h\equiv r \pmod 2}^{r-1} \zeta_h^+ + \zeta_{r,i}^+.$$

A similar analysis, but simpler, gives

$$\zeta_{p,i} \cong \zeta_{i,p} = \sum_{h=0,\ h\not\equiv r \pmod 2}^{r-1} \zeta_h^+.$$

On the other hand, $S_p \otimes S_p$ is the direct sum of the space $(S_p \otimes S_p)^s$ of symmetric tensors of degree 2 over S_p and of the space $(S_p \otimes S_p)^a$ of alternating tensors; $(S_p \otimes S_p)^s$ is spanned by the elements $u \otimes v + v \otimes u$, $u, v \in S_p$, and $(S_p \otimes S_p)^a$ by the elements $u \otimes v - v \otimes u$. We have a similar decomposition for $S_i \otimes S_i$. Let $\zeta_{pp}^s, \zeta_{pp}^a, \zeta_{ii}^s, \zeta_{ii}^a$ be the representations of G_+ on the spaces $(S_p \otimes S_p)^s$, $(S_p \otimes S_p)^a$, $(S_i \otimes S_i)^s$, $(S_i \otimes S_i)^a$. Taking formula (2) into account, we obtain

$$\zeta_{pp}^s \cong \sum_{h=0,\ h\equiv r(4)}^{r-1} \zeta_h^+ + \zeta_{rp}^+,$$

$$\zeta_{pp}^a \cong \sum_{h=0,\ h\equiv r+2(4)}^{r-1} \zeta_h^+,$$

$$\zeta^s_{ii} \cong \sum_{h=0,\ h\equiv r(4)}^{r-1} \zeta_h^+ + \zeta_{rr}^+,$$

$$\zeta_{ii}^a \cong \sum_{h=0,\ h\equiv r+2(4)}^{r-1} \zeta_h^+.$$

III.4.5. *Let $(\xi_1, \cdots, \xi_r)$ be a base of an even (respectively: odd) maximal totally singular space Z. Then $\xi_1 \wedge \cdots \wedge \xi_r$ is in $E_{r,p}$ (respectively: $E_{r,i}$) and $E_{r,p}$ is spanned by elements of this form.*

Let u be a representative spinor for Z; then we have $u \in S_p$ if Z is even, $u \in S_i$ if Z is odd, and

$$\beta_r(u, u) = uf\alpha(u) = a(\xi_1 \wedge \cdots \wedge \xi_r),$$

a a scalar $\neq 0$, which proves the first assertion. The elements of the form $\xi_1 \wedge \cdots \wedge \xi_r$ are obviously permuted among themselves by the operations of $\zeta_r(G)$; since $\zeta_{rp}^+, \zeta_{ri}^+$ are simple, this proves the second assertion.

3.5. Imbedded Spaces

III.5.1. *Let M' be a nonisotropic $(m - 2)$-dimensional subspace of M, Q' the restriction of Q to M' and Γ'^{+} the special Clifford group of Q'. Then either one of the half-spin representations of Γ^{+} induces a representation of Γ'^{+} which is equivalent to the spin representation.*

We know that the representation of Γ'^{+} induced by a half-spin representation of Γ^{+} is the sum of a certain number of representations of Γ'^{+}, which are all equivalent to the spin representation (II.6.2). This representation is of degree 2^{r-1}. Since the Clifford algebra C' of Q' is of dimension $2^{2(r-1)}$, its simple representations are of degree $\equiv 0 \pmod{2^{r-1}}$, and the spin representation of Γ'^{+} can occur only once in the representation induced by a half-spin representation of Γ^{+}. The argument also proves that C' is isomorphic to a full matrix algebra over K.

Let us now consider the case where Q' is itself of maximal index $r - 1$. Let $M' = N' + P'$ be a representation of M' as the sum of two totally singular subspaces N' and P' of dimension $r - 1$. Then N' is contained in at least one maximal totally singular subspace N_1 of M and P' in exactly two maximal totally singular subspaces P_1, P_2 of M, one of which is even and the other odd (III.1.11). Since $N' \cap P' = \{0\}$, $P_i \cap N_1$ is of dimension ≤ 1; $P_1 \cap N_1$ and $P_2 \cap N_1$ cannot both be of dimension 1 in virtue of III.1.10; assume then that $P_1 \cap N_1 = \{0\}$, whence $M = N_1 + P_1$. We shall assume that N_1 and P_1 are the spaces N and P which we have selected for the study of Q. Let $C^{N'}$ be the subalgebra of C (or C') generated by N'; then we may take $C^{N'}$ to be the space of spinors S' for Q'. We propose to define explicitly an isomorphism of S' with $C_+^{N} = S_p$, which realizes the equivalence of the representation of Γ'^{+} induced by ρ_p^{+} with the spin representation of Γ'^{+}. The restrictions of B to $N' \times P'$ and to $N \times P$ being nondegenerate, it is clear that we can find vectors $x_0 \in N$, $y_0 \in P$ with the following properties: x_0 and y_0 are in the conjugate of $N' + P'$, and $B(x_0, y_0) = 1$. We then have $N = N' + Kx_0$, $P = P' + Ky_0$. Set $C_+^{N'} = C^{N'} \cap C_+$, $C_-^{N'} = C^{N'} \cap C_-$; then $C^{N'}$ is the direct sum of these two spaces, which are the spaces of even and odd half-spinors for Q'. Define a linear mapping φ of $C^{N'}$ into S_p by the formula

$$\varphi(u'_+ + u'_-) = u'_+ + u'_- x_0 \qquad (u'_+ \in C_+^{N'}, u'_- \in C_-^{N'}).$$

Let ρ' be the spin representation of C' and ρ_p^{+} the half-spin representation of C_+ on S_p. We propose to compute $\varphi(\rho'(v') \cdot u')$ for $u' \in S'$, $v' \in C'$. Let f' be the product of the elements of a base of P'. We have by definition $(\rho'(v')\cdot u')f' = v'u'f'$. On the other hand, it is clear that $f'y_0$ is the product of the elements of a base of P and therefore differs only by a scalar factor from f, whence $(\rho'(v')\cdot u')f = v'u'f$. Now, x_0 is orthogonal to every element of P, whence $x_0 f = (-1)^r f x_0$, and it follows that $(\rho'(v')\cdot u')x_0 f = v'u'x_0 f$. If we decompose u' in $u'_+ + u'_-$, $u'_+ \,\varepsilon\, C_+^{N'}$, $u'_- \,\varepsilon\, C_-^{N'}$, we have

$$\varphi(\rho'(v')\cdot u') = \rho'(v')\cdot u'_+ + (\rho'(v')\cdot u'_-)x_0 \qquad \text{if} \qquad v' \,\varepsilon\, C'_+,$$

$$\varphi(\rho'(v')\cdot u') = (\rho'(v')\cdot u'_+)x_0 + \rho'(v')\cdot u'_- \qquad \text{if} \qquad v' \,\varepsilon\, C'_-.$$

If $v' \,\varepsilon\, C'_+ \subset C_+$, then we have $(\rho^+{}_p(v)\cdot u)f = vuf$ for all $u \,\varepsilon\, S_p$. It follows immediately that

$$\varphi(\rho'(v')\cdot u') = \rho^+{}_p(v')\cdot\varphi(u') \qquad \text{if} \qquad v' \,\varepsilon\, C'_+.$$

If $v' \,\varepsilon\, C'_-$, $\rho^+{}_p(v')$ is not defined because v' is not in C_+. In order to treat that case, we set $\xi_0 = x_0 + y_0$, whence $\xi_0^2 = 1$. On the other hand, we observe that the center of C'_+ contains an element z' of square 1 which anticommutes with every element of C'_- (II.2.4; if K is of characteristic 2, we take $z' = 1$). If K is of characteristic 2, then $z'f = f$. If not, the simple ideals of which C'_+ is the sum are $C'_+(1 - z')$ and $C'_+(1 + z')$. Since f' is a half-spinor for Q' and the ideals $C'_+(1 - z')$ and $C'_+(1 + z')$ are the kernels of the two half-spin representations of C'_+, one of the elements $(1 - z')f'$, $(1 + z')f'$ is 0; replacing if necessary z' by $-z'$, we may assume that $z'f' = f'$, whence $z'f = f$. This being said, we have

$$\varphi(\rho'(v')\cdot u')f = v'u'_+x_0 f + v'u'_- f,$$

$$(\rho^+{}_p(v'\xi_0 z')\cdot\varphi(u'))f = v'\xi_0 z'u'_+ f + v'\xi_0 z'u'_- x_0 f.$$

Since ξ_0 is in the conjugate space of M', it anticommutes with every element of M'; it follows that $\xi_0 z'$ commutes with every element of M', and therefore also of C'. Thus, we have $\xi_0 z'u'_+ f = u'_+\xi_0 z'f = u'_+\xi_0 f$. Similarly, x_0 anticommutes with every element of M' and commutes with z', whence $v'\xi_0 z'u'_- x_0 f = v'u'_-\xi_0 x_0 f$. It is clear that $y_0 f = 0$, whence $\xi_0 f = x_0 f$. We have $x_0^2 = 0$, $x_0 y_0 + y_0 x_0 = 1$, whence $\xi_0 x_0 f = y_0 x_0 f = f$. We conclude that

$$\varphi(\rho'(v')\cdot u') = \rho^+{}_p(v'\xi_0 z')\cdot\varphi(u') \qquad (v' \,\varepsilon\, C'_-).$$

Let ψ be the mapping of C' into C_+ defined by

$$\psi(v'_+ + v'_-) = v_+' + v'_-\xi_0 z' \qquad \text{if} \qquad v'_+ \,\varepsilon\, C'_+, v'_- \,\varepsilon\, C'_-.$$

Since $(\xi_0 z')^2 = \xi_0^2 z'^2 = 1$ and $\xi_0 z'$ commutes with every element of

C', ψ is clearly a homomorphism, and therefore an isomorphism, since C' is simple. We have

$$\varphi(\rho'(v')\cdot u') = \rho_p^{+}(\psi(v'))\cdot\varphi(u') \qquad (v' \in C').$$

The mapping $\rho_p^{+} \circ \psi$ is a representation of degree 2^{r-1} of C', and therefore equivalent to ρ'. Since ρ' is simple and $\varphi \neq 0$, φ is an isomorphism of S' with S_p by Schur's lemma. If $s' \in \Gamma'^{+}$, then

$$\varphi(\rho'(s')\cdot u') = \rho_p^{+}(s')\cdot\varphi(u'),$$

and φ realizes the equivalence of ρ' and of the representation of Γ'^{+} induced by ρ_p^{+}. Moreover, we have determined explicitly a representation $\rho_p^{+} \circ \psi$ of C' on S_p which extends the restriction of ρ_p^{+} to C_+'.

Let now S_i be the space of odd half-spinors for Q; define a linear mapping φ' of S' into S_i by

$$\varphi'(u_+' + u_-') = u_+'x_0 + u_-' \qquad (u_+' \in C_+^{N'},\ u_-' \in C_-^{N'}).$$

Then it is easily seen that φ' is a linear isomorphism and that

$$\varphi'(\rho'(v')\cdot u') = \rho_i^{+}(\psi(v'))\cdot\varphi'(u') \qquad (v' \in C_+').$$

We shall now determine the elements $u' \in S'$ such that $\varphi(u')$ is a pure spinor for Q. Let Z be any even maximal totally singular subspace of M, u a representative spinor for Z, and u' the element of S' such that $\varphi(u') = u$. The space $Z \cap M'$ is of dimension $r - 1$ or $r - 2$ (for M' cannot contain any totally singular space of dimension r). If $x' \in Z \cap M'$, then we have $\varphi(\rho'(x')\cdot u') = \rho_p^{+}(\psi(x'))\cdot u$, $\psi(x') = x'\xi_0 z' = \xi_0 z' x'$, and $\rho_p^{+}(\psi(x')) = \rho(\xi_0 z')\rho(x')$, where ρ is the spin representation of C. Since $x' \in Z$, we have $\rho(x')\cdot u = 0$, whence $\rho'(x')\cdot u' = 0$. If $Z \cap M'$ is of dimension $r - 1$, it is a maximal totally singular subspace of M', and u' is a representative spinor for this space (III.1.4). Conversely, any maximal totally singular subspace of M' is contained in exactly one even maximal totally singular subspace of M (III.1.11), which shows that the image under φ of any pure spinor for Q' is a pure spinor for Q.

Assume now that $\dim (Z \cap M') = r - 2$. Then $Z \cap M'$ is contained in exactly two maximal totally singular subspaces Z_+', Z_-' of M', with Z_+' even and Z_-' odd (III.1.4); let u_+', u_-' be representative spinors for Z_+', Z_-'. Then $u_+ = \varphi(u_+')$ and $u_- = \varphi(u_-')$ are pure spinors for Q and represent even maximal totally singular subspaces Z_+, Z_- of M such that $Z_+ \cap M' = Z_+'$, $Z_- \cap M' = Z_-'$. The space $Z_+ \cap Z_-$ contains $Z \cap M'$, which is of dimension $r - 2$; since Z_+, Z_- are distinct, it follows from III.1.10 that $Z_+ \cap Z_- = Z \cap M'$. Making use of III.1.12, we see that u is a linear combination of u_+ and u_-, and therefore that u' is a linear combination of u_+' and u_-'.

Let conversely u'_+, u'_- be pure spinors for Q', representative for maximal totally singular subspaces Z'_+, Z'_- of M' such that dim $(Z'_+ \cap Z'_-) = r - 2$. Then the same argument as above shows that $\varphi(u'_+)$, $\varphi(u'_-)$ are representative spinors for even maximal totally singular subspaces of M whose intersection is of dimension $r - 2$; therefore, $\varphi(u'_+ + u'_-)$ is pure for Q in virtue of III.1.12. Thus, we have the following results:

III.5.2. *Let M' be a nonisotropic $(m - 2)$-dimensional subspace of M (with $m > 2$), Q' the restriction of Q to M', which we assume to be of index $r - 1$, S' the space of spinors for Q', S_p the space of even half-spinors for Q and φ the isomorphism of S' with S_p constructed above. Let u' be in S'; in order for $\varphi(u')$ to be pure for Q, it is necessary and sufficient that one of the following conditions be satisfied: (a) u' is pure for Q'; or (b) $u' = u_+' + u_-'$, where u_+' and u_-' are pure for Q' and represent maximal totally singular subspaces Z_+', Z_-' of M' whose intersection is of dimension $r - 2$. In case (a), $\varphi(u')$ represents a maximal totally singular subspace Z of M whose intersection with M' is of dimension $r - 1$ and represented by u'; in case (b), $\varphi(u')$ represents a maximal totally singular subspace Z of M whose intersection with M' is $Z_+' \cap Z_-'$.*

Let $S = S_p + S_i$ be the space of spinors for Q. Let e' be the product of the elements of a base of N'; then $e = e'x_0$ is the product of the elements of a base of N. We have associated to e a bilinear invariant β of the spin representation of Γ_0^+ and to e' a bilinear invariant β' of the spin representation of $\Gamma_0'^+$ (Section 3.2). We shall now investigate the mutual relationship between β, β' and the mappings φ, φ' introduced above. Let u' and v' be elements of S', with $u' = u'_+ + u'_-$, $v' = v'_+ + v'_-$, u'_+, v'_+ in $C_+^{N'}$, u'_-, v'_- in $C_-^{N'}$. If u, v are in S, $\beta(u, v)$ is defined by the condition that $\beta(u, v)e$ is the homogeneous component of degree r of $\alpha(u)v$. We have

$$\alpha(\varphi(u'))\varphi(v') = (\alpha(u'_+) + x_0\alpha(u'_-))(v'_+ + v'_-x_0).$$

An element of $C^{N'}$ has 0 as its homogeneous component of degree r. Since $e'x_0 = e$ and x_0 anticommutes with every element of $C_-^{N'}$ and commutes with every element of $C_+^{N'}$, we have

$$\beta(\varphi(u'), \varphi(v')) = \beta'(u'_+, v'_-) - \beta'(u'_-, v'_+).$$

A similar computation gives

$$\beta(\varphi'(u'), \varphi'(v')) = \beta'(u'_-, v'_+) - \beta'(u'_+, v'_-),$$

$$\beta(\varphi(u'), \varphi'(v')) = \beta'(u'_+, v'_+) + \beta'(u'_-, v'_-).$$

3.6. The Kernels of the Half-Spin Representations

III.6.1. *Assume that the dimension m of M is ≥ 6. Then the kernels of the half-spin representations of Γ^+ are of order 1 if K is of characteristic 2, of order 2 if K is of characteristic $\neq 2$. In the latter case, these kernels are $\{1, z\}$ and $\{1, -z\}$, where z is an element of the center of C_+ whose square is 1 and which anticommutes with every element of M.*

Let s be an element of the kernel of the half-spin representation ρ_p^+ of Γ^+ on the even half-spinors, and let $\sigma = \chi(s)$ be the image of s under the vector representation. If Z is any even maximal totally singular subspace of M and u a representative spinor for Z, then $\rho(s)\cdot u = u$; but $\rho(s)\cdot u$ is a representative spinor for $\sigma(Z)$, whence $\sigma(Z) = Z$. Now, let x be any singular vector $\neq 0$ in M. We shall see that, under the assumption $m \geq 6$, Kx is the intersection of all even maximal totally singular spaces which contain it. The space Kx is contained in at least one maximal totally singular space, and therefore also, since $r \geq 3$, in at least one totally singular space of dimension $r - 1$. Making use of III.1.11, we see that Kx is contained in some even maximal totally singular space Z_1. Let x_1 be an element of Z_1 not in Kx. Since $r \geq 3$, x belongs to some $(r - 2)$-dimensional subspace U of Z_1 which does not contain x_1. The space $U + Kx_1$, of dimension $r - 1$, is contained in some odd maximal totally singular space Z'. Since x_1 is not in U, there is a subspace V of dimension $r - 1$ of Z' which contains U but not x_1, and V is contained in some even maximal totally singular space Z_2. We have $U \subset Z_2$, whence $Kx \subset Z_2$; but $Z_2 \cap Z'$ is V, to which x_1 does not belong, and x_1 is not in Z_2. This shows that Kx is the intersection of all even maximal totally singular spaces which contain it, whence $\sigma(Kx) = Kx$. Thus, we have $\sigma\cdot x = a(x)x$ for any singular vector x, $a(x)$ being a scalar. If x and y are singular, orthogonal to each other and linearly independent, then $x + y$ is singular, and $\sigma\cdot(x + y) = a(x + y)\,(x + y)$, whence $a(x) = a(y) = a(x + y)$. It follows immediately that there exist scalars a and b such that $\sigma\cdot x = ax$ for all $x \in N$, $\sigma\cdot y = by$ for all $y \in P$. If x is an element $\neq 0$ in N, there is at least a $y \neq 0$ in P such that $B(x, y) = 0$, since $r \geq 2$. This shows that $a = b$ and that $\sigma\cdot x = ax$ for all $x \in M$ (since $M = N + P$). Since σ belongs to the orthogonal group of Q, we have $a^2 = 1$. If K is of characteristic 2, then $a = 1$, and s is in the center of C, whence $s = c\cdot 1$, $c \in K$. Since $\rho_p^+(s)$ is the identity, $c = 1$ and $s = 1$. Assume now that K is not of characteristic 2. If $a = 1$, we see as above that $s = 1$. If $a = -1$, we observe that the center of C_+ contains an element z of square 1 which anticommutes with every element of M and that the simple ideals of which C_+ is the sum are $C_+ (1 - z)$ and $C_+ (1 + z)$ (II.2.4). One at least of z, $-z$ is therefore in the kernel of ρ_p^+, and we may assume that it is z. If $s' = sz^{-1}$, then s' is in the kernel of ρ_p^+ and $\chi(s')$ is the identity, whence $s' = 1$, $s = z$.

The assertion relative to the kernel of ρ^+_i may be proved exactly in the same manner; III.6.1 is thereby proved.

III.6.2. *Let M_1 be an even-dimensional space over a field K and Q_1 a quadratic form on M_1 whose associated bilinear form is nondegenerate. Assume that the algebra $(C_1)_+$ of even elements of the Clifford algebra C_1 of Q_1 is not simple. Let Γ^+_1 be the special Clifford group of Q_1. Assume that* dim $M_1 \geq 6$. *Then, if K is of characteristic 2, the half-spin representations of Γ^+_1 are faithful. If K is of characteristic $\neq 2$, then the kernels of these representations are $\{1, z\}$ and $\{1, -z\}$, where z is an element of the center of $(C_1)_+$ whose square is 1 and which anticommutes with every element of M_1.*

If K is not of characteristic 2, we know that $(C_1)_+$ contains an element z with the stated properties. Let K' be an algebraically closed overfield of K. Since any quadratic form on a finite-dimensional space over K' is of maximal index (when its associated bilinear form is nondegenerate), III.6.2 follows from III.6.1 and from what has been said in Section 2.7.

3.7. The Case $m = 6$

We shall assume in this section that M is of dimension 6, *except in the statement of* III.7.3.

Let (x_1, x_2, x_3) and (y_1, y_2, y_3) be bases of N and P such that $B(x_i, y_j) = \delta_{ij}$ $(1 \leq i, j \leq 3)$, $y_1y_2y_3 = f$. We set $u_0 = e = x_1x_2x_3$, $u_i = x_i$ $(i = 1, 2, 3)$. Then u_0, u_1, u_2, u_3 form a base of S_i. Every element $u \neq 0$ of S_i is a pure spinor. This is clear if $u \in Ku_0$. If not, then $u = x + ce$, $x \in N$, $x \neq 0$, $c \in K$ and we may write $ce = xyz$, where $y, z \in N$, whence $u = x \exp (yz)$, which shows that u is pure (III.1.9). It follows immediately that every element $\neq 0$ of S_p is likewise a pure spinor.

III.7.1. *The representation ρ^+_i of Γ^+ on the space S_i maps Γ^+ onto the group of all automorphisms of S_i whose determinants are squares of elements of K.*

We first prove that

$$\det \rho^+_i(s) = \lambda^2(s)$$

if $s \in \Gamma^+$. First let s be in Γ^+_0; then $\chi(s)$ is in G^+_0, which is the commutator subgroup of G^+ (II.3.9). Since $G^+ = \chi(\Gamma^+)$, the commutator subgroup of G^+ is the image of that of Γ^+ under χ, and there is an element s' of the commutator subgroup $(\Gamma^+)'$ of Γ^+ such that $\chi(s') = \chi(s)$, whence $s' = cs$, $c \in K$. It is clear that $\lambda((\Gamma^+)') = \{1\}$, whence $\lambda(s') = 1 = \lambda(s)$ and $c^2 = 1$, $c = \pm 1$. On the other hand, $\rho_i^+(s') = c\rho_i^+(s)$, whence $\det \rho_i^+(s) = c^{-4} \det \rho_i^+(s')$. But we have $\det \rho_i^+(s') = 1$, since $s' \in (\Gamma^+)'$, whence $\det \rho_i^+(s) = 1$. Now, let s be any element of Γ^+. Let k be an element $\neq -1$ in K; set $t = 1 + ky_1x_1$. It is clear that t commutes with x_2, x_3, y_2, y_3. We have $\alpha(t) = 1 + kx_1y_1$, $\alpha(t)t = (1 + k)\cdot 1$, as follows immediately from the fact that $x_1y_1 + y_1x_1 = 1$, $x_1^2 = y_1^2 = 0$. Thus, we have

$$t^{-1} = (1 + k)^{-1}(1 + kx_1y_1),$$

and

$$tx_1t^{-1} = (1 + k)^{-1}x_1 .$$

Writing

$$t = 1 + k - kx_1y_1 ,$$

$$t^{-1} = (1 + k)^{-1}(1 + k - ky_1x_1),$$

we see that

$$ty_1t^{-1} = (1 + k)y_1 .$$

Thus, t is in Γ, and obviously in Γ^+, and $\lambda(t) = 1 + k$. Let us now compute $\rho_i(t)$. The operation $\rho(y_1)$ is an antiderivation which maps x_1 upon 1, x_2, x_3 upon 0. Thus, we have $\rho(x_1y_1)\cdot u_0 = u_0$, $\rho(x_1y_1)\cdot u_1 = u_1$, $\rho(x_1y_1)\cdot u_2 = \rho(x_1y_1)\cdot u_3 = 0$ and $\rho(t)\cdot u_0 = u_0$, $\rho(t)\cdot u_1 = u_1$, $\rho(t)\cdot u_2 = (1 + k)u_2$, $\rho(t)\cdot u_3 = (1 + k)u_3$, whence $\det \rho_i^+(t) = (1 + k)^2 = \lambda^2(t)$. Select k in such a way that $(1 + k)\lambda(s) = 1$: then $ts \in \Gamma_0^+$, and $\det \rho_i^+(s) = (\det \rho_i^+(t))^{-1} = (1 + k)^{-2} = \lambda^2(s)$, which proves our formula.

This being said, let μ be any automorphism of S_i whose determinant is a square. Then $\mu(u_0)$ is representative for an odd maximal totally singular space Z. There is a $\sigma_1 \in G^+$ such that $\sigma_1(Z) = N$; write $\sigma_1 = \chi(s_1)$, $s_1 \in \Gamma^+$, whence $\rho_i^+(s)\cdot\mu(u_0) = au_0$, $a \in K$. We set $\mu_1 = \rho_i^+(s_1)\mu$. We have $N \subset S_i$; let x be $\neq 0$ in N and let s be an operation of Γ^+ such that $\chi(s)$ transforms N into itself. Then $\rho_i^+(s)\cdot x$ is representative for a maximal totally singular space Z'_x such that $\chi(s)\cdot x \in Z'_x$, as follows immediately from the fact that x is representative for a space Z_x such that $x \in Z_x$ (III.1.4). This being said, we may write $\mu_1(x) = f(x) + a(x)u_0$, f being a linear mapping of N into itself and $a(x)$ a scalar. Since μ_1 transforms Ku_0 into itself, f is an automorphism of N. Since

N is totally singular, f is a Q-automorphism and may therefore be extended to an operation $\sigma_2 \in G$; since $\sigma_2(N) = N$, σ_2 is in G^+ and we may write $\sigma_2 = \chi(s_2)$, $s_2 \in \Gamma^+$. For any $x \in N$, $\rho_i^+(s_2)\cdot x$ is representative for a maximal totally singular space Z'_x such that $f(x) \in Z'_x$. It follows that

$$\rho_i^+(s_2)\cdot x = c(x)(\exp v(x))f(x), \qquad c(x) \in K, \qquad v(x) \in C_2^N,$$

i.e., $\rho_i^+(s_2)x = c(x)f(x) + a'(x)u_0$. If $k \in K$, then clearly, we have $c(kx) = c(x)$, since $f(kx) = kf(x)$. If x, y are linearly independent, then we have

$$\begin{aligned} c(x + y)f(x + y) &= c(x)f(x) + c(y)f(y) \\ &= c(x + y)(f(x) + f(y)), \end{aligned}$$

whence $c(x) = c(y) = c(x + y)$. It follows that $c(x)$ is constant on the set of element $x \neq 0$ of N; let c be its constant value and $s_3 = c^{-1}s_2$: then $\mu_1(x) \equiv \rho_i^+(s_3)\cdot x \pmod{Ku_0}$ for all $x \in N$, and $\rho_i^+(s_3)u_0 \in Ku_0$. Let $\mu_2 = \rho_i^+(s_3^{-1})\mu_1$: then $\mu_2(x) = x + a_1(x)u_0$ for $x \in N$, a_1 being a linear function on N. Now, observe that, if $v = c_1x_2x_3 + c_2x_3x_1 + c_1x_1x_2$, then $(\exp v)x_i = c_iu_0$ $(i = 1, 2, 3)$, $(\exp v)u_0 = u_0$. Take $c_i = -a_1(x_i)$ $(i = 1, 2, 3)$, and set $s_4 = \exp v$. Then $\rho_i^+(s_4)$ is the operation of multiplication by $\exp v$ in S_i; if $\mu_2 = \rho_i^+(s_4)\mu_3$, then we have $\mu_3(x) = x$ for $x \in N$, $\mu_3(u_0) = bu_0$, $b \in K$. We have $\det \mu_3 = b$; since μ_3 is the product of μ_1 by an element of $\rho_i^+(\Gamma^+)$, b is a square. For any $d \neq 0$ in K, we have constructed above an element $t = t_1$ of Γ^+ such that $\rho_i^+(t_1)$ changes u_0 into itself, u_1 into u_1, u_2 into du_2, u_3 into du_3, and we have $\lambda(t_1) = d$. We may similarly construct elements t_i $(i = 2, 3)$ such that $\lambda(t_i) = d$, and $\rho_i^+(t_i)$ changes u_0 and u_i into themselves, u_j into du_j if $j \neq 0, i$. Then $\rho_i^+(d^{-2}t_1t_2t_3)$ changes u_i into u_i $(i = 1, 2, 3)$, u_0 into $d^{-2}u_0$. If we select d so that $d^{-2} = b$, we have $\mu_3 = \rho_i^+(d^{-2}t_1t_2t_3)$, which concludes the proof of III.7.1.

III.7.2. *The group $\rho_i^+(\Gamma_0^+)$ is the group of automorphisms of determinant* 1 *of S_i.*

We know already that $\det \rho_i^+(s) = 1$ if $s \in \Gamma_0^+$ (proof of III.7.1). Let μ be an automorphism of determinant 1 of S_i; then we may write $\mu = \rho_i^+(s)$, $s \in \Gamma^+$, and $\lambda^2(s) = \det \mu = 1$, whence $\lambda(s) = \pm 1$. If $\lambda(s) \neq 1$, then K is not of characteristic 2 and the kernel of ρ_i^+ contains an element z of the center of C_+ such that $z^2 = 1$. If $(\xi_1, \cdots, \xi_6)$ is a base of M composed of mutually orthogonal vectors, then z is a scalar multiple of $\xi_1 \cdots \xi_6$, from which it follows easily that $\alpha(z) = -z$, whence $\lambda(z) = -1$; we then have $\mu = \rho_i^+(sz)$, $\lambda(sz) = 1$.

It follows immediately from the preceding results that $\rho_{\bullet}^{+}(\Gamma^{+})$ (respectively: $\rho^{+}_{\bullet}(\Gamma^{+}_{0})$) is the group of automorphisms of $S_{\bullet}$ whose determinants are squares in K (respectively: are 1).

Now, let M' be the conjugate of the space spanned by x_1, y_1. Then M' is not isotropic and of dimension 4; the restriction Q' of Q to M' is of index 2. Let Γ'^{+} and $\Gamma_0'^{+}$ be the special Clifford group and the restricted Clifford group of Q'. Identifying the Clifford algebra C' of Q' to the subalgebra of C generated by M', Γ'^{+} and $\Gamma_0'^{+}$ are subgroups of Γ^{+} and Γ_0^{+}, respectively; the representation ρ_i^{+} of Γ^{+} induces a representation of Γ'^{+} which is equivalent to the spin representation of this group (see II.6.2). Let S'_i be the subspace of S_i spanned by u_0, u_1 and S''_i the subspace spanned by u_2, u_3; we shall see that these spaces are invariant by the operations of $\rho_i^{+}(\Gamma'^{+})$. The space S'_i is the set of elements $u \in S_i$ such that $\rho(x_1)\cdot u = x_1 u = 0$; since $\rho(y_1)$ is an antiderivation which maps any $x \in N$ upon $B(x, y_1)\cdot 1$, S''_i is the space of elements $u \in S_i$ such that $\rho(y_1)\cdot u = 0$. We have, for any $s \in \Gamma^{+}$,

$$\rho_i^{+}(sx_1s^{-1})\cdot(\rho_i^{+}(s)\cdot u) = \rho_i^{+}(s)\cdot(\rho_i^{+}(x_1)\cdot u),$$

and a similar formula for y_1; it follows immediately that $\rho^{+}_i(s)$ maps S'_i and S''_i into themselves if $s \in \Gamma'^{+}$. If we denote by $\rho_i'^{+}(s)$, $\rho_i''^{+}(s)$ the restrictions of $\rho_i^{+}(s)$ to S'_i, S''_i, then $\rho_i'^{+}$ and $\rho_i''^{+}$ are equivalent to the two half-spin representations of Γ'^{+}. We shall see that the determinants of $\rho_i'^{+}(s)$, $\rho_i''^{+}(s)$ are both equal to $\lambda(s)$. This is obvious if K has only 2 elements. If not, then the reduced orthogonal group $G_0'^{+}$ of Q' is the commutator subgroup of its special orthogonal group (II.3.9), and we see exactly as in the proof of III.7.1 that, if $s \in \Gamma_0'^{+}$, then $\det \rho_i'^{+}(s) = \det \rho_i''^{+}(s) = 1$. On the other hand, if we set $t_2 = 1 + ky_2x_2$, with $k \neq -1$, then we see as in the proof of III.7.1 that $t_2 \in \Gamma^{+}$, that $\lambda(t_2) = 1 + k$, and that $\rho_i^{+}(t_2)$ transforms u_0 into u_0, u_1 into $(1 + k)u_1$, u_2 into u_2, and u_3 into $(1 + k)u_3$. It is clear that $t_2 \in \Gamma'^{+}$ and that $\det \rho_i'^{+}(t_2) = \det \rho_i''^{+}(t_2) = \lambda(t_2)$. If $s \in \Gamma'^{+}$, then we can determine k in such a way that $\lambda(t_2)\lambda(s) = 1$, whence $\det \rho_i'^{+}(t_2s) = \det \rho_i''^{+}(t_2s) = 1$, $\lambda(t_2s) = 1$, which proves our assertion.

Conversely, let μ be an automorphism of S such that $\mu(S'_i) = S'_i$, $\mu(S''_i) = S''_i$ with the property that the determinants of the restrictions of μ to S'_i, S''_i are equal to each other. Then $\det \mu$ is a square, and there is an s in Γ^{+} such that $\mu = \rho_i^{+}(s)$. Then $\rho^{+}(\chi(s)\cdot x_1)$ maps the elements of S'_i upon 0. This shows that $\chi(s)\cdot x_1$ belongs to the maximal totally singular spaces whose representative spinors are x_1 and $x_1x_2x_3$, i.e., $\chi(s)x_1 \in Kx_1$. A similar argument shows that $\chi(s)\cdot y_1 \in Ky_1$. Since $B(\chi(s)\cdot x_1, \chi(s)\cdot y_1) = 1$, we have $\chi(s)\cdot x_1 = cx_1$, $\chi(s)\cdot y_1 =$

$c^{-1}y_1$ for some $c \in K$. Let $k = c - 1$, $t = 1 + ky_1x_1$; then (see proof of III.7.1), $\chi(t^{-1}s)$ leaves x_1 and y_1 fixed, whence $\chi(t^{-1}s) \in G'^+$ and $t^{-1}s = cs'$, with some $s' \in \Gamma'^+$. The determinants of the restrictions of $\rho_i^+(t)$ to S'_i, S''_i are 1 and $(1 + k)^2 = c^2$, respectively. By our condition on μ, these determinants are equal to each other, whence $c = \pm 1$. If $c \neq 1$, we observe that the kernel of ρ_i^+ contains an element z which anticommutes with every element of M, and $\mu = \rho_i^+(zs)$, so that we are reduced to the case where $c = 1$, in which case $t = 1$ and $s = s' \in \Gamma'^+$.

In particular, we see that $\rho_i^+(\Gamma_0'^+)$ is the group of automorphisms of S_i which leave S'_i, S''_i invariant and whose restrictions to these spaces are of determinant 1. This gives the following results:

III.7.3. *If $m = 4$, then Γ_0^+ is the direct product of two subgroups each one of which is isomorphic to the group of automorphisms of determinant 1 of a 2-dimensional vector space over K. These groups are the kernels of the two half-spin representations of Γ_0^+.*

3.8. The Case of Odd Dimension

We denote by $\overline{M}$ a vector space of odd dimension $2r - 1$ over a field K of characteristic $\neq 0$ and by $\overline{Q}$ a quadratic form on $\overline{M}$ whose associated bilinear form $\overline{B}$ is nondegenerate and which is of maximal index $r - 1$. We denote by $\overline{C}$ the Clifford algebra of $\overline{Q}$, by $\overline{C}_+$ the algebra of even elements of $\overline{C}$, by $\overline{S}$ the space of spinors for $\overline{Q}$, by ρ^+ the spin representation of $\overline{C}_+$, by $\overline{\Gamma}$, $\overline{\Gamma}^+$, $\overline{\Gamma}_0^+$ the group of Clifford, the special Clifford group, and the reduced Clifford group of $\overline{Q}$, and by $\bar{\rho}$, $\bar{\rho}^+$, $\bar{\rho}_0^+$ the spin representations of these groups.

We select two maximal totally singular subspaces N', P' of $\overline{M}$ whose sum $M' = N' + P'$ is direct and a nonisotropic subspace of $\overline{M}$. We select a basic vector ξ_0 of the conjugate space of M', and we set

$$a = Q(\xi_0).$$

We may imbed $\overline{M}$ in a vector space M of dimension $m = 2r$ which is the sum of $\overline{M}$ and of a one-dimensional space spanned by a vector ξ_0'; we extend $\overline{Q}$ to a quadratic form Q on M by setting

$$Q(\bar{x} + c\xi_0') = \overline{Q}(\bar{x}) - ac^2 \qquad (c \in K)$$

if $\bar{x} \in \overline{M}$. It is then clear that Q is of rank $2r$. It is furthermore of index r, for $Q(\xi_0 + \xi'_0) = 0$ and $\xi_0 + \xi'_0$ is orthogonal to every element of N' (relatively to the associated bilinear form B of Q), which shows that $N = N' + K(\xi_0 + \xi'_0)$ is totally singular for Q. Moreover, $P = P' + K(\xi_0 - \xi'_0)$ is totally singular and $M = N + P$. We shall use for Q the notation which was introduced earlier in this chapter, it being understood that N and P are the spaces which were used in defining the space of spinors.

Either one of the half-spin representations of Γ^+ induces a representation of degree 2^{r-1} of $\overline{\Gamma}^+$; this representation is equivalent to the spin representation of this group in virtue of II.6.2. Let $\rho_p{}^+$ be the representation of Γ^+ on the even half-spinors. Denote by $\zeta_h{}^+$ the representation of G^+ on the h-vectors, and by $\zeta_{rp}{}^+$ the one of the two simple representations of which $\zeta_r{}^+$ is the sum which occurs in $\rho_p{}^+ \otimes \rho_p{}^+$. Denote by $\theta_h{}^+$ the representation $s \rightarrow \lambda(s)\zeta_h{}^+(\chi(s))$ of Γ^+ (where $\lambda(s)$ is the norm of s) and by $\theta_{rp}{}^+$ the representation $s \rightarrow \lambda(s)\zeta_{rp}{}^+(\chi(s))$. Then we have proved in Section 3.4 that

$$\rho_p{}^+ \otimes \rho_p{}^+ \cong \sum_{h=0, h\equiv r(2)}^{r-2} \theta_h{}^+ + \theta_{rp}{}^+.$$

It follows that $\bar{\rho}^+ \otimes \bar{\rho}^+$ is the sum of representations respectively equivalent to the representations of $\overline{\Gamma}^+$ induced by the $\theta_h{}^+(0 \leq h \leq r - 2,\ h \equiv r \pmod 2)$ and $\theta_{rp}{}^+$. In order to study these representations, denote by $\overline{E}$ the exterior algebra of $\overline{M}$, which we identify with the subalgebra of the exterior algebra E on M which is generated by $\overline{M}$. Let E_h and $\overline{E}_h$ be the spaces of homogeneous elements of degree h of E, $\overline{E}$, whence $\overline{E}_h = \overline{E} \cap E_h$. Taking a base of M composed of a base of $\overline{M}$ and of ξ'_0, we see immediately that every element of E_h is uniquely representable in the form $u \wedge \xi'_0 + v$, where $u \in \overline{E}_{h-1}$ and $v \in \overline{E}_h$. We identify the orthogonal group $\overline{G}$ of $\overline{Q}$ to the subgroup of G composed of the operations which leave ξ'_0 fixed, and we denote by $\bar{\zeta}_h{}^+$ the representation of $\overline{G}^+$ on the h-vectors. It is clear that, if $\bar{\sigma} \in \overline{G}^+$, then

$$\zeta_h{}^+(\bar{\sigma})\cdot(u \wedge \xi'_0 + v) = (\bar{\zeta}_{h-1}{}^+(\bar{\sigma})\cdot u) \wedge \xi'_0 + \bar{\zeta}_h{}^+(\bar{\sigma})\cdot v$$

if $u \in \overline{E}_{h-1}$, $v \in \overline{E}_h$. This shows that the representation of $\overline{G}^+$ induced by $\zeta_h{}^+$ is equivalent to $\bar{\zeta}_{h-1}{}^+ + \bar{\zeta}_h{}^+$ if $h > 0$, to $\bar{\zeta}_0{}^+$ if $h = 0$. This applies in particular if $h = r$; in that case, $\bar{\zeta}_{r-1}{}^+$ and $\bar{\zeta}_r{}^+$ are equivalent to each other (because $r + (r - 1)$ is the dimension of $\overline{M}$) and are simple. Since $\zeta_r{}^+$ is equivalent to the sum of $\zeta_{rp}{}^+$ and of another representation $\zeta_{ri}{}^+$ of the same degree as $\zeta_{rp}{}^+$, it follows immediately that the representation of $\overline{G}^+$ induced by $\zeta_{rp}{}^+$ is equivalent to $\bar{\zeta}_{r-1}{}^+$. Let $\bar{\theta}_h$ be the representation $s \rightarrow \lambda(s)\bar{\zeta}_h{}^+(\bar{\chi}(s))$ of $\overline{\Gamma}^+$; then we obtain the formula

$$\bar{\rho}^+ \otimes \bar{\rho}^+ \cong \sum_{h=0}^{r-1} \bar{\theta}_h{}^+. \tag{1}$$

Let $\bar{\varphi}_p$ be an isomorphism of $\overline{S}$ with the space S_p of even half-spinors for Q such that

$$\rho_p{}^+(\bar{s}) \circ \bar{\varphi}_p = \bar{\varphi}_p \circ \bar{\rho}^+(\bar{s})$$

for all $\bar{s} \in \bar{\Gamma}^+$. Let S be the space of spinors for Q. Then we have defined for each h $(0 \leq h \leq 2r)$ a bilinear mapping β_h of $S_p \times S_p$ into E_h such that

$$\beta_h(\rho(s)\cdot u,\ \rho(s)\cdot v) = \lambda(s)\zeta_h(\chi(s))\cdot\beta_h(u, v)$$

for all $s \in \Gamma$, $u, v \in S$, ρ being the spin representation of Γ (see Section 3.4); β_h is identically zero on $S_p \times S_p$ if $h \not\equiv r \pmod 2$ but not if $h \equiv r \pmod 2$. If $0 \leq h \leq 2r - 1$, let h' be equal to h if $h \equiv r \pmod 2$, to $h + 1$ if $h \not\equiv r \pmod 2$. Let $\bar{u}, \bar{v}$ be in $\bar{S}$; then $\beta_{h'}(\bar{\varphi}_p(\bar{u}), \bar{\varphi}_p(\bar{v}))$ may be represented in the form $\bar{w} \wedge \xi'_0 + \bar{w}'$, where $\bar{w} \in \bar{E}_{h'-1}$ ($\bar{w} = 0$ if $h' = 0$), $\bar{w}' \in \bar{E}_h$. We define $\bar{\beta}_h(\bar{u}, \bar{v})$ to be $\bar{w}$ if $h' = h + 1$, $\bar{w}'$ if $h' = h$. Thus, $\bar{\beta}_h$ is a bilinear mapping of $\bar{S} \times \bar{S}$ into $\bar{E}_h$, and we have

$$\bar{\beta}_h(\bar{\rho}^+(\bar{s})\cdot\bar{u},\ \bar{\rho}^+(\bar{s})\cdot\bar{v}) = \lambda(\bar{s})\bar{\zeta}_h^{\,+}(\bar{\chi}(\bar{s}))\cdot\bar{\beta}_h(\bar{u}, \bar{v})$$

for any $\bar{s} \in \bar{\Gamma}^+$, $\bar{u} \in \bar{S}$, $\bar{v} \in \bar{S}$. It is easily seen that $\bar{\beta}_h \neq 0$ for $0 \leq h \leq 2r - 1$. It is easy to verify that

$$\bar{\beta}_h(\bar{v}, \bar{u}) = (-1)^{(r-h)(r-h-1)/2}\bar{\beta}_h(\bar{u}, \bar{v})$$

(see formula (2), Section 3.4). If we set

$$\bar{\beta}_0(\bar{u}, \bar{v}) = \bar{\beta}(\bar{u}, \bar{v})\cdot 1, \qquad \bar{\beta}(\bar{u}, \bar{v}) \in K,$$

$\bar{\beta}$ is a bilinear form on $\bar{S} \times \bar{S}$, which is an invariant of the spin representation of $\bar{\Gamma}^+_0$.

The mappings $\bar{\beta}_h$, $\bar{\beta}$ depend on the choice of the isomorphism $\bar{\varphi}_p$. It should be observed, however, that this mapping is determined up to a scalar factor. For any other isomorphism $\bar{\varphi}'_p$ with the same property as $\bar{\varphi}_p$ is of the form $\bar{\varphi}_p \circ \omega$, where ω is an automorphism of the vector space $\bar{S}$ which commutes with every operation of $\bar{\rho}^+(\bar{\Gamma}^+)$. Since $\bar{\Gamma}^+$ is a set of generators of the algebra $\bar{C}_+$ (II.4.2), ω commutes with every operation of $\bar{\rho}^+(\bar{C}_+)$. But $\bar{\rho}^+(\bar{C}_+)$ is of dimension 2^{2r-2}, and $\bar{S}$ of dimension 2^{r-1}, which shows that $\bar{\rho}^+(\bar{C}_+)$ is the algebra of all endomorphisms of $\bar{S}$ and therefore that ω is a scalar multiple of the identity.

We shall now extend to the odd-dimensional case the notion of a pure spinor. Let $\bar{Z}$ be a maximal totally singular subspace of $\bar{M}$; then $\bar{Z}$ is of dimension $r - 1$ and therefore contained in a uniquely determined even maximal totally singular subspace Z of M (III.1.11). Let u be a representative spinor for Z, and let $\bar{u}$ be the element of $\bar{S}$ such that $\bar{\varphi}_p(\bar{u}) = u$. Then $\bar{u}$ depends on the choice of $\bar{\varphi}_p$, but the one-dimensional space $K\bar{u}$ depends only on $\bar{Z}$. Any basic element of this space is called a *representative spinor* for $\bar{Z}$; any element of $\bar{S}$ which is representative for some maximal totally singular space is called a *pure spinor*. If Z is any maximal totally singular subspace of M, then $Z \cap \overline{M}$ is of dimension $r - 1$ or r and cannot be of dimension r, since it is totally singular. It follows that *a necessary and sufficient condition for a spinor* $\bar{u} \in \bar{S}$ *to be pure is for* $\bar{\varphi}_p(\bar{u})$ *to be pure.*

III.8.1. *Let* $\bar{Z}$ *be a maximal totally singular subspace of* $\overline{M}$ *and* $\bar{u}$ *a representative spinor for* $\bar{Z}$. *Let* $\bar{z}$ *be an odd invertible element of the center of* $\bar{C}$; *then* $\bar{Z}$ *is the set of all elements* $\bar{x} \in \overline{M}$ *such that* $\bar{\rho}^+(\bar{z}\bar{x}) \cdot \bar{u} = 0$ *and any element of* $\bar{S}$ *which is mapped upon* 0 *by all* $\bar{\rho}^+(\bar{z}\bar{x})$, $\bar{x} \in \bar{Z}$, *is in* $K\bar{u}$. *If* $\bar{s} \in \bar{\Gamma}^+$, *then* $\bar{\rho}^+(\bar{s}) \cdot \bar{u}$ *is representative for* $\bar{\chi}(\bar{s})(\bar{Z})$.

Let $u = \bar{\varphi}_p(\bar{u})$; then u is a pure spinor for Q, which is representative for the even maximal totally singular subspace Z of M containing $\bar{Z}$. If $\bar{x} \in \overline{M}$, then we have $\bar{\varphi}_p(\bar{\rho}^+(\bar{z}\bar{x}) \cdot \bar{u}) = \rho_p{}^+(\bar{z}\bar{x}) \cdot \bar{\varphi}_p(\bar{u}) = \rho(\bar{z}) \cdot (\rho(\bar{x}) \cdot \bar{\varphi}_p(\bar{u}))$. Since $\bar{z}$ is invertible, a necessary and sufficient condition for $\bar{x}$ to be in $\bar{Z}$, or what amounts to the same, in Z is that $\bar{\rho}^+(\bar{z}\bar{x}) \cdot \bar{u} = 0$ (see III.1.4). The space $\bar{Z}$ is also contained in some odd maximal totally singular subspace Z' of M; let u' be a representative spinor for Z'. We shall see that $Ku + Ku'$ is the space of all spinors for Q which are mapped upon 0 by all $\rho(\bar{x})$, $\bar{x} \in \bar{Z}$. We can find an operation σ of the orthogonal group G of Q which transforms $\bar{Z}$ into a subspace Z_1 of N. Let $(x_1, \cdots, x_{r-1})$ be a base of Z_1. Let u'' be a spinor such that $\rho(\bar{x}) \cdot u'' = 0$ for all $\bar{x} \in \bar{Z}$. If we set $u''_1 = \rho(s) \cdot u''$, then we have $x_i u''_1 = \rho(x_i) u''_1 = 0$ $(1 \le i \le r - 1)$, which means that the element u''_1 of the exterior algebra C^N of N is a multiple of $x_1 \cdots x_{r-1}$. If x_r is an element of N not in Z_1, then u''_1 is a linear combination of $x_1 \cdots x_{r-1}$ and $x_1 \cdots x_{r-1}x_r$, which are representative spinors for the two maximal totally singular subspaces of M containing Z_1. It follows that u'' is a linear combination of u and u'. Now, if $\bar{u}''$ is an element of $\bar{S}$ such that $\bar{\rho}^+(\bar{z}\bar{x}) \cdot \bar{u}'' = 0$ for all $\bar{x} \in \bar{Z}$, and $u'' = \bar{\varphi}_p(\bar{u}'')$, then u'' is a linear combination of u, u' by what we have just said. But u, u'' are even half-spinors and u' an odd half-spinor, which shows that $u'' \in Ku$ and $\bar{u}'' \in K\bar{u}$. If $\bar{s} \in \bar{\Gamma}^+ \subset \Gamma^+$, then $\rho_p{}^+(\bar{s}) \cdot u$ is representative for $\chi(\bar{s})(Z)$, whose intersection with $\overline{M}$ is $\bar{\chi}(\bar{s})(\bar{Z})$, which shows that $\bar{\rho}^+(\bar{s}) \cdot \bar{u}$ is representative for $\bar{\chi}(\bar{s})(\bar{Z})$.

III.8.2. *A necessary and sufficient condition for a spinor* $\bar{u} \neq 0$ *in* $\bar{S}$ *to be pure is that* $\bar{\beta}_h(\bar{u}, \bar{u}) = 0$ *for* $0 \le h < r - 1$.

Let $u = \bar{\varphi}(\bar{u})$; a necessary and sufficient condition for u to be pure is that $\beta_h(u, u) = 0$ for $k \neq r$ (III.4.4). Since u is even, it is already sufficient that $\beta_k(u, u) = 0$ for $0 \le k < r$ by III.4.3. On the other hand, $\beta_k(u, u)$ is always 0 if $k \not\equiv r \pmod 2$ by III.4.2. Our assertion then follows immediately from the definition of the mappings $\bar{\beta}_h$.

From now on, we shall assume that the Clifford algebra $\bar{C}$ of $\bar{Q}$ is not simple. Let $(x_1, \cdots, x_{r-1})$ and $(y_1, \cdots, y_{r-1})$ be bases of N' and P' such that $\bar{B}(x_i, y_j) = \delta_{ij}$ $(1 \leq i, j \leq r-1)$. Then $(x_1, \cdots, x_{r-1}, y_1, \cdots, y_{r-1}, \xi_0)$ is a base of $\bar{M}$ and the discriminant of $\bar{B}$ with respect to this base is $2a(-1)^{r-1}$. Making use of II.2.6, we see that a must then be a square in K. Thus, under proper choice of ξ_0, we may assume that $a = 1$. Let Q' be the restriction of $\bar{Q}$ to $M' = N' + P'$, and C' be the Clifford algebra of Q'. The center of the algebra C'_+ of even elements of C' is spanned by 1 and by an element z' of square 1 which anticommutes with every element of M'; z' may be selected in such a way that $z'f = f$ (see Section 3.5). If we set $\bar{z} = z'\xi_0$, then $\bar{z}$ is an odd invertible element of the center of $\bar{C}$ and $\bar{z}^2 = 1$. In Section 3.5 we have constructed an isomorphism ψ of C' with a subalgebra of C_+:

$$\psi(u'_+ + u'_-) = u'_+ + u'_-z'\xi_0$$

if $u'_+ \in C'_+$, $u'_- \in C'_-$. Since $z'\xi_0 = \bar{z}$, we see that $\psi(C') \subset \bar{C}_+$. Since C' and $\bar{C}_+$ are of the same dimension 2^{2r-2}, we have $\psi(C') = \bar{C}_+$. The reciprocal mapping of ψ' is an isomorphism of $\bar{C}_+$ with C'. This isomorphism may be extended to a homomorphism π of $\bar{C}$ onto C'. For any element of $\bar{C}$ may be uniquely represented in the form $\bar{u}_+ + \bar{u}_-$, where $\bar{u}_+ \in \bar{C}_+$, $\bar{u}_- \in \bar{C}_-$, and $\bar{u}_+ + \bar{u}_- \to \bar{u}_+ + \bar{u}_-\bar{z}$ is a homomorphism of $\bar{C}$ onto $\bar{C}_+$. Composing this homomorphism with the reciprocal of ψ, we obtain a mapping π with the required property. If $x' \in M'$, then we have $x' = (x'\bar{z})\bar{z}$, whence $\pi(x') = x'$; as for ξ_0, we write also $\xi_0 = (\xi_0\bar{z})\bar{z} = z'\bar{z}$, and we see that $\pi(\xi_0) = z'$; π induces an isomorphism of $\bar{M} = M' + K\xi_0$ with the subspace $\bar{M}' = M' + Kz'$ of C'. If $\bar{x} \in \bar{M}$, then we have $(\pi(\bar{x}))^2 = \pi(\bar{x}^2) = Q(\bar{x})\cdot 1$. If $\bar{s} \in \bar{\Gamma}$, then we have $\bar{s}\bar{M}\bar{s}^{-1} = \bar{M}$, which shows that $\pi(\bar{s})\bar{M}'(\pi(\bar{s}))^{-1} = \bar{M}'$. Conversely, let s' be an invertible element of C' such that $s'\bar{M}'s'^{-1} = \bar{M}'$. Then s' is the image under π of some element $\bar{s} \in \bar{C}_+$ and, since $\pi(\bar{z}) = 1$, $\bar{M}\bar{z} \subset \bar{C}$, we have $\bar{s}(\bar{M}\bar{z})\bar{s}^{-1} = \bar{M}$, whence $\bar{s}\bar{M}\bar{s}^{-1} = \bar{M}$ and $\bar{s} \in \bar{\Gamma}^+$. Thus, we see that π induces a homomorphism of the group $\bar{\Gamma}$ onto the group $\bar{\Gamma}'$ of invertible elements $s' \in C'$ such that $s'\bar{M}'s'^{-1} = \bar{M}'$; $\pi(\bar{\Gamma})$ is identical to $\pi(\bar{\Gamma}^+)$, and π induces an isomorphism on $\bar{\Gamma}^+$.

The group $\bar{\Gamma}^+$ is generated by the products $\bar{x}\bar{z}$, where $\bar{x}$ runs over the invertible elements of $\bar{M}$ (II.3.4). It follows that the group $\bar{\Gamma}'$ is generated by the invertible elements of $\bar{M}'$ (i.e., by the elements of this space whose squares are not 0). If ρ' is the spin representation of C', then $\rho' \circ \pi$ induces a representation of $\bar{C}_+$ on the space S' of spinors for Q', and this representation is obviously equivalent to the spin representation of $\bar{C}_+$. Thus, $\rho' \circ \pi$ induces a representation of $\bar{\Gamma}^+$ on S' which is equivalent to the spin representation of $\bar{\Gamma}^+$. The norm homomorphism $\bar{\lambda}$ of $\bar{\Gamma}^+$ defines a homomorphism $\bar{\lambda}'$ of $\bar{\Gamma}'$ into the multiplicative group of elements $\neq 0$ in K. We shall say that $\bar{\lambda}'$ is the *norm homomorphism* of $\bar{\Gamma}'$.

If $(\xi_1, \cdots, \xi_m)$ is a base of $\bar{M}$ composed of mutually orthogonal vectors, $\xi_1 \cdots \xi_m$ is an odd invertible element of the center of $\bar{C}$ and is therefore a scalar multiple of $\bar{z}$. It follows immediately that $\bar{\alpha}(\bar{z}) = (-1)^{r-1}\bar{z}$. If $x' \in M'$, then we have $\alpha'(x') = x'$, $\bar{\alpha}(\psi(x')) = \bar{\alpha}(x'\bar{z}) = \bar{\alpha}(\bar{z})x' = (-1)^{r-1}\, \psi(x')$. It follows that $\bar{\alpha} \circ \psi$ coincides with $\psi \circ \alpha'$ on C'_+, with $(-1)^{r-1}\, \psi \circ \alpha'$ on C'_-. If we denote by $\tilde{\alpha}'$ the product of the main involution of C' by its main antiautomorphism, then we see that $\bar{\alpha} \circ \psi$ coincides with $\psi \circ \alpha'$ if r is odd, with $\psi \circ \tilde{\alpha}'$ if r is even. In particular, if we denote by $\bar{\Gamma}'_0$ the kernel of the norm homomorphism of $\bar{\Gamma}'$, then we have

$$\bar{\Gamma}'_0 \cap \Gamma'^+ = \Gamma_0'^+.$$

Making use of the remark at the end of Section 3.2, we see that, if $\bar{s}' \in \bar{\Gamma}'$, $u', v' \in S'$, then

$$\beta'(\rho'(\bar{s}')\cdot u', \rho'(\bar{s}')\cdot v') = \bar{\lambda}'(\bar{s}')\beta'(u', v')$$

if r is odd, while

$$\tilde{\beta}'(\rho'(\bar{s}')\cdot u', \rho'(\bar{s}')\cdot v') = \bar{\lambda}'(\bar{s}')\tilde{\beta}'(u', v')$$

if r is even; in these formulas, β' and $\tilde{\beta}'$ are the bilinear forms on $S' \times S'$ which were introduced in Section 3.2.

CHAPTER IV

THE PRINCIPLE OF TRIALITY

We shall denote by M an 8-dimensional vector space over a field K and by Q a quadratic form on M of rank 8, of defect 0 (in case K is of characteristic 2) and of index 4. We shall use the notation introduced in Chapter III; in particular, we denote by N and P two four-dimensional totally singular subspaces of M which are supplementary to each other, by f the product in the Clifford algebra C of Q of the elements of a base of P, and by C^N the subalgebra of C generated by N; we take $S = C^N$ to be the space of spinors for Q, and we denote by S_p and S_i the spaces of even and odd half-spinors. The representations of the subgroup Γ^+ of the group of Clifford on the spaces S, S_p , S_i are denoted by ρ^+, ρ_p^+, ρ_i^+; the vector representation of the group of Clifford Γ is denoted by χ, and its spin representation by ρ.

We have constructed in Section 3.2 a bilinear form β on $S \times S$, defined as follows: if u, v are in S, then $\beta(u, v)e$ is the homogeneous component of degree 4 of $\alpha(u)v$, where α is the main antiautomorphism of C. Since $r = 4$, β is symmetric and vanishes on $S_p \times S_i$ and on $S_i \times S_p$ and its restrictions to $S_p \times S_p$ and $S_i \times S_i$ are nondegenerate. If $z \in M$, then we have

$$\beta(\rho(z) \cdot u,\ \rho(z) \cdot v) = Q(z)\beta(u, v),$$

and, for any $s \in \Gamma$,

$$\beta(\rho(s) \cdot u,\ \rho(s) \cdot v) = \lambda(s)\beta(u, v).$$

Moreover, since $r = 4$, there exists a quadratic form γ on S such that

$$\gamma(u + v) = \gamma(u) + \gamma(v) + \beta(u, v)$$

for any u, $v \in S$, and

$$\gamma(\rho(z) \cdot u) = Q(z)\gamma(u) \qquad (z \in M),$$

$$\gamma(\rho(s) \cdot u) = \lambda(s)\gamma(u) \qquad (s \in \Gamma).$$

This form has been explicitly constructed; if K is not of characteristic 2, then we have $\gamma(u) = \frac{1}{2}\beta(u, u)$.

4.1. A New Characterization of Pure Spinors

IV.1.1. *Let u be an element $\neq 0$ of S. In order for u to be a pure spinor, it is necessary and sufficient that the following conditions be satisfied: u is either even or odd, and $\gamma(u) = 0$.*

Suppose first that u is a representative spinor for a maximal totally singular space Z. There is an operation σ of G which transforms Z into P, whence $\rho(s)\cdot u = a\cdot 1$, a a scalar $\neq 0$, since 1 is a representative spinor for P. But $\gamma(1)$ is obviously 0, whence $\gamma(u) = (\lambda(s))^{-1}\,\gamma(\rho(s)\cdot u) = 0$. Assume now that our conditions are satisfied. Proceeding exactly as in the proof of III.3.2, we see that there is an $s \in \Gamma$ such that $\rho(s)\cdot u$ is even, and the homogeneous component of degree 0 of $\rho(s)\cdot u$ (in C^N, identified to the exterior algebra of N) is $\neq 0$, while its homogeneous component of degree 2 is 0. Since N is of dimension 4, $\rho(s)\cdot u$ is then of the form $\alpha\cdot 1 + be$, $\alpha \neq 0$ (where e is the product of the elements of a base of N). If K is not of characteristic 2, then $\beta(\rho(s)\cdot u, \rho(s)\cdot u)\cdot e$ is the homogeneous component of degree 4 of

$$\alpha(\rho(s)\cdot u)\rho(s)\cdot u = (a\cdot 1 + be)^2,$$

whence

$$\gamma(\rho(s)\cdot u) = \tfrac{1}{2}\beta(\rho(s)\cdot u, \rho(s)\cdot u) = ab.$$

The formula $\gamma(\rho(s)\cdot u) = ab$ is also true in case K is of characteristic 2, in view of our explicit construction of γ (Section 3.2). Since $\gamma(\rho(s)\cdot u) = \lambda(s)\gamma(u) = 0$, we have $b = 0$ and $\rho(s)\cdot u = a\cdot 1$ is pure, which shows that u is pure.

4.2. Construction of an Algebra

We shall now introduce the vector space $A = M \times S$, of dimension $8 + 16 = 24$. This space is the direct sum of the two subspaces $M \times \{0\}$ and $\{0\} \times S$; we shall identify these two spaces to M and S, respectively. It should be observed that the spaces M and S, as they have been defined, have the space N in common. Our identification is therefore logically illicit; in all rigor, we should consider A as the sum of two spaces respectively isomorphic to M and to S. We do not do it, in order to avoid complications of notation, but it should be kept in mind that the elements of N are now doubled: they appear either as elements of M or as elements of S and should be distinguished from each other according as to whether they function in one or the other capacity.

We define a quadratic form Ω on A by the formula

$$\Omega(x + u) = Q(x) + \gamma(u) \qquad (x \in M,\ u \in S).$$

The bilinear form associated with Ω will be denoted by Λ. If $x, x' \in M$, $u, u' \in S$, then we have

$$\Lambda(x + u', x' + u') = B(x, x') + \beta(u, u').$$

It follows immediately that Λ is nondegenerate. The subspaces M, S_p, S_i of A are nonisotropic (with respect to Λ), and the conjugate of any one of them is the sum of the other two.

We shall now define a cubic form F on A by the formula

$$F(x + u + u') = \beta(\rho(x)\cdot u, u'),$$

where $x \in M$, $u \in S_p$, $u' \in S_i$. We have, by III.2.2,

$$F(x + u + u') = \beta(\rho(x)\cdot u, u') = \beta(u, \rho(x)\cdot u'). \tag{1}$$

From the cubic form F we deduce by the process of polarization a trilinear form Φ on the space $A \times A \times A$. The form Φ is defined as follows. Let ξ, η, ζ be the elements of A. Then

$$\Phi(\xi, \eta, \zeta) = F(\xi + \eta + \zeta) + F(\xi) + F(\eta) + F(\zeta) - [F(\xi + \eta) + F(\eta + \zeta) + F(\zeta + \xi)].$$

Let $(\xi_1, \cdots, \xi_{24})$ be a base of A, and set

$$\xi = \sum_{i=1}^{24} a_i\xi_i, \qquad \eta = \sum_{i=1}^{24} b_i\xi_i, \qquad \zeta = \sum_{i=1}^{24} c_i\xi_i.$$

Then it is easily verified that $\Phi(\xi, \eta, \zeta)$ is of the form

$$\sum_{i,j,k=1}^{24} d_{ijk}a_ib_jc_k,$$

where the d_{ijk}'s are fixed constants, which proves that Φ is trilinear. If K has infinitely many elements, then $\Phi(\xi, \eta, \zeta)$ may also be defined as follows: $F(a\xi + b\eta + c\zeta)$ being expressed as a polynomial in a, b, c, then the coefficient of the term in abc in this polynomial is $\Phi(\xi, \eta, \zeta)$. The trilinear form Φ is obviously symmetric.

We can now define a law of composition in A. Let ξ and η be in A. Then $\zeta \to \Phi(\xi, \eta, \zeta)$ is a linear form on A. Since the bilinear form Λ is nondegenerate, there exists a unique element ω of A such that $\Lambda(\omega, \zeta) = \Phi(\xi, \eta, \zeta)$ for all $\zeta \in A$. We set $\omega = \xi \circ \eta$; thus, we have

$$\Phi(\xi, \eta, \zeta) = \Lambda(\xi \circ \eta, \zeta).$$

The mapping $(\xi, \eta) \to \xi \circ \eta$ is obviously bilinear; it is the law of composition of a structure of (nonassociative) algebra on A. Since Φ is symmetric, this algebra is obviously commutative.

The law of composition in the algebra A being defined in terms of the forms Ω and F only, it is clear that any automorphism of the vector space A which leaves these two forms invariant is an automorphism of the algebra A.

IV.2.1. *If each one of the elements ξ, η, ζ lies in one of the spaces M, S_p, S_i, then we have*

$$\Phi(\xi, \eta, \zeta) = F(\xi + \eta + \zeta).$$

We have $F(\omega) = 0$ if ω lies in one of the three spaces $M + S_p$, $S_p + S_i$, $S_i + M$, as follows immediately from the definition of F. Thus, under our assumption, $F(\xi)$, $F(\eta)$, $F(\zeta)$, $F(\xi + \eta)$, $F(\eta + \zeta)$, $F(\zeta + \xi)$ are all zero, which proves IV.2.1.

Suppose that ξ and η are both in M, or both in S_p or both in S_i. Then we have $\Phi(\xi, \eta, \zeta) = 0$ if ζ lies in either one of the spaces M, S_p, S_i, and therefore, by linearity, $\Phi(\xi, \eta, \zeta) = 0$ for any $\zeta \in A$. This proves.

IV.2.2. *We have $\xi \circ \eta = 0$ if ξ, η both lie in one and the same of the spaces M, S_p, S_i.*

Now, assume that $\xi \in M$, $\eta \in S_p$. Then we have $\Phi(\xi, \eta, \zeta) = 0$ when $\zeta \in M + S_p$, whence $\Lambda(\xi \circ \eta, \zeta) = 0$ for $\zeta \in M + S_p$. Thus, $\xi \circ \eta$ lies in the conjugate of $M + S_p$ with respect to Λ, i.e., in S_i. Proceeding in a similar manner, we obtain the formulas

$$M \circ S_p \subset S_i \qquad S_p \circ S_i \subset M \qquad S_i \circ M \subset S_p. \tag{2}$$

IV.2.3. *Let x be in M and u in S. Then we have*

$$x \circ u = \rho(x)\cdot u \qquad \gamma(x \circ u) = Q(x)\gamma(u) \qquad x \circ (x \circ u) = Q(x)u.$$

It is clearly sufficient to prove the first of these formulas in the case where u is in either S_p or S_i. Assume for instance that $u \in S_p$, whence $x \circ u \in S_i$. If $u' \in S_i$, we have $\beta(x \circ u, u') = \Lambda(x \circ u, u') = \Phi(x, u, u') = F(x + u + u') = \beta(\rho(x)\cdot u, u')$, whence $x \circ u = \rho(x)\cdot u$, since the restriction of β to $S_i \times S_i$ is nondegenerate. We proceed in exactly the same way if $u \in S_i$. The second formula of IV.2.3 follows immediately from the first. Moreover, we have $x \circ u \in S$, whence $x \circ (x \circ u) = \rho(x)\cdot(\rho(x)\cdot u) = \rho(x^2)\cdot u = Q(x)u$, since $x^2 = Q(x)\cdot 1$.

If x, y are in M and $u \in S$, then we have

$$\beta(x \circ u, y \circ u) = B(x, y)\gamma(u) \tag{3}$$

The left side is $\gamma((x + y) \circ u) - \gamma(x \circ u) - \gamma(y \circ u)$, which is $(Q(x + y) - Q(x) - Q(y))\gamma(u)$ by IV.2.3, which proves (3).

Now, let s be an element of Γ. If $x \in M$, $u \in S_p$, $u' \in S_i$, set

$$\mu(s)\cdot(x + u + u') = \chi(s)\cdot x + \rho(s)\cdot u + \rho(s)\cdot u'.$$

Then μ is clearly a linear representation of the group Γ. We have $Q(\chi(s)\cdot x) = Q(x)$, $\gamma(\rho(s)\cdot(u + u')) = \lambda(s)\gamma(u + u')$. If $s \in \Gamma_0$, then $\mu(s)$ leaves the quadratic form Ω invariant. We shall now prove the formula

$$F(\mu(s)\cdot\omega) = \lambda(s)F(\omega) \qquad (s \in \Gamma, \omega \in A). \tag{4}$$

Set $\omega = x + u + u'$, $x \in M$, $u \in S_p$, $u' \in S_i$. We have

$$F(\mu(s)\cdot\omega) = \beta(\rho(\chi(s)\cdot x)\cdot(\rho(s)\cdot u), \rho(s)\cdot u').$$

Now, $\chi(s)x = sxs^{-1}$, whence

$$\rho(\chi(s)\cdot x) = \rho(s)\rho(x)(\rho(s))^{-1}$$

and

$$\begin{aligned} F(\mu(s)\cdot\omega) &= \beta(\rho(s)\rho(x)\cdot u, \rho(s)\cdot u') \\ &= \lambda(s)\beta(\rho(x)\cdot u, u') \\ &= \lambda(s)F(\omega), \end{aligned}$$

which proves (4). It follows immediately that

$$\Phi(\rho(s)\cdot\xi, \rho(s)\cdot\eta, \rho(s)\cdot\zeta) = \lambda(s)\Phi(\xi, \eta, \zeta) \tag{5}$$

if ξ, η, ζ are in A.

Let x be in M and u, u' in S. Then

$$\left.\begin{aligned} \rho(s)\cdot(x \circ u) &= \mu(s)\cdot(x \circ u) \\ &= \mu(s)\cdot x \circ \mu(s)\cdot u \\ &= \chi(s)\cdot x \circ \rho(s)\cdot u, \\ (\mu(s)\cdot u) \circ (\mu(s)\cdot u') &= (\rho(s)\cdot u) \circ (\rho(s)\cdot u') \\ &= \lambda(s)(\chi(s)\cdot(u \circ u')) \\ &= \lambda(s)\cdot(\mu(s)\cdot(u \circ u')). \end{aligned}\right\} \tag{6}$$

For we have

$$\begin{aligned} \chi(s)\cdot x \circ \rho(s)\cdot u &= \rho(\chi(s)\cdot x)\rho(s)\cdot u \\ &= \rho(s)\rho(x)\cdot u, \end{aligned}$$

which proves the first formula. We have $\rho(s)\cdot u \circ \rho(s)\cdot u' \in M$, and, for $y \in M$,

$$B(\rho(s)\cdot u \circ \rho(s)\cdot u', y) = \Phi(\mu(s)\cdot u, \mu(s)\cdot u', y)$$
$$= \lambda(s)\Phi(u, u', \mu(s^{-1})\cdot y)$$
$$= \lambda(s)B(u \circ u', \chi(s^{-1})\cdot y)$$

and

$$B(\chi(s)\cdot u \circ u', y) = B(u \circ u', \chi(s^{-1})\cdot y),$$

which proves the second formula (6).

It is clear that every automorphism of the vector space A which leaves the quadratic form Ω and the cubic form F invariant leaves also the trilinear form Φ invariant and is therefore an automorphism of the structure of algebra of A. Thus, every operation of $\mu(\Gamma_0)$ is an automorphism of the algebra A.

IV.2.4. *Any automorphism σ of the algebra A which transforms each one of the spaces M and S into itself belongs to the group $\mu(\Gamma_0)$.*

We know that the representation ρ of the Clifford algebra C maps C onto the algebra of all endomorphisms of the vector space S. Thus, there exists an element $s \in C$ such that $\sigma\cdot u = \rho(s)\cdot u$ for all $u \in S$. Since σ induces an automorphisms of S, s is invertible in C. Let x be in M; then we have $\rho(s)\rho(x)\cdot u = \sigma(x \circ u) = \sigma\cdot x \circ \sigma\cdot u = \rho(\sigma\cdot x)\rho(s)\cdot u$, and $\rho(\sigma\cdot x) = \rho(s)\rho(x)\rho(s^{-1}) = \rho(sxs^{-1})$. Since ρ is a faithful representation of C, $sxs^{-1} = \sigma\cdot x$, which proves that $s \in \Gamma$, and that $\sigma = \mu(s)$. It remains only to prove that $\lambda(s) = 1$. We have, for $u, u' \in S$, $\mu(s)\cdot u \circ u' = \mu(s)\cdot u \circ \mu(s)\cdot u'$. Comparing with the second formula of (6), we see that $\lambda(s) = 1$ provided there exist elements $u, u' \in S$ such that $u \circ u' \neq 0$. Now, let x_0 be a nonsingular vector in M and u' an element $\neq 0$ of S_i. Since $(\rho(x_0))^2\cdot u = Q(x_0)u$, the mapping $u \longrightarrow \rho(x_0)\cdot u$ of S_p into S_i is one-to-one; since S_p and S_i have the same dimension, it is an isomorphism of S_p with S_i, and there exists an $u \in S_p$ such that $\beta(\rho(x_0)\cdot u, u') \neq 0$; since $B(u \circ u', x_0) = \beta(\rho(x_0)\cdot u, u_i)$ (by IV.2.1), we have $u \circ u' \neq 0$ and IV.2.4 is proved.

4.3. The Principle of Triality

IV.3.1 (Principle of triality). *There exists an automorphism J of order 3 of the vector space A which has the following properties: J leaves the quadratic form Ω and the cubic form F invariant; J maps M onto S_p, S_p onto S_i, and S_i onto M.*

There exists an element $x_1 \varepsilon M$ such that $Q(x_1) = 1$. For, if x and y are elements of M such that $Q(x) = Q(y) = 0$, $B(x, y) = 1$, then $x_1 = x + y$ has the required property. We have $x_1 \varepsilon \Gamma$ and $\lambda(x_1) = x_1^2 = 1$, whence $x_1 \varepsilon \Gamma_0$. It follows that the operation $\mu(x_1)$ (see Section 4.2) leaves Ω and F invariant; it is clear that $\mu(x_1)$ maps M onto itself, S_p onto S_i, and S_i onto S_p.

Now, we know that the restrictions of γ to S_p and S_i are of index 4. If u, v are in S_p, $\gamma(u) = \gamma(v) = 0$, and, if $\beta(u, v) = 1$, then we have $\gamma(u + v) = 1$. We see in the same way that γ takes the value 1 at some point of S_i.

Now, let u_1 be any point of S_p such that $\gamma(u_1) = 1$. We shall associate to u_1 an automorphism τ of A. If $x \varepsilon M$, then we set $\tau(x) = u_1 \circ x = x \circ u_1 \varepsilon S_i$. If $x, y \varepsilon M$, then we have

$$\beta(\tau \cdot x, \tau \cdot y) = B(x, y)$$

by formula (3), Section 4.2. Thus, if $\tau \cdot y = 0$, then $B(x, y) = 0$ for all $x \varepsilon M$ and $y = 0$; this proves that $x \to \tau \cdot x$ is a one-to-one mapping of M into S_i. Since M and S_i have the same dimension 8, our mapping is a linear isomorphism of M with S_i. Every $u' \varepsilon S_i$ may therefore be written in one and only one way in the form $\tau \cdot x$, for some $x \varepsilon M$, and we set $\tau \cdot u' = x$. Having now defined τ on M and S_i, we extend it by linearity to $M + S_i$; we obtain in this way an automorphism of order 2 of $M + S_i$. There remains to define τ on S_p, which we do by the formula

$$\tau \cdot u = \beta(u, u_1)u_1 - u \qquad (u \varepsilon S_p). \tag{1}$$

If τ' is the symmetry in S_p with respect to the conjugate hyperplane of Ku_1, relative to the restriction of the quadratic form γ to S_p, then we have $\tau \cdot u = -\tau' \cdot u$ (since $\gamma(u_1) = 1$). It follows that the mapping of S_p into itself defined by (1) is an automorphism of order 2 of S_p. Completing the definition of τ by linearity, we see that we obtain an automorphism τ of order 2 of A which maps any $x \varepsilon M$ upon $u_1 \circ x$. This automorphism leaves the forms Ω and F invariant. For, let x be in M, u in S_p, and u' in S_i. Then we have $\tau \cdot x \varepsilon S_i$, $\tau \cdot u \varepsilon S_p$, $\tau \cdot u' \varepsilon M$, and

$$\Omega(\tau \cdot (x + u + u')) = \gamma(u_1 \circ x) + \gamma(\tau \cdot u) + Q(\tau \cdot u').$$

We have $\gamma(u_1 \circ x) = \gamma(u_1)Q(x)$ by IV. 2.3, and this is $Q(x)$. We have $\gamma(\tau \cdot u) = \gamma(u)$ because the restriction of τ to S_p belongs to the orthogonal group of the restriction of γ to S_p. If $u' = u_1 \circ y$, with $y \varepsilon M$, then we have $\tau \cdot u' = y$, whence $\gamma(u') = Q(y) = \gamma(\tau \cdot u')$, and this shows that τ leaves Ω invariant. We have

$$F(\tau \cdot (x + u + u')) = \beta(\tau \cdot u, \rho(\tau \cdot u') \circ (u_1 \circ x))$$

by formula (1), Section 4.2. Again let $u' = u_1 \circ y = \rho(y) \cdot u_1$; since $u_1 \circ x = \rho(x) \cdot u_1$, we have $\rho(\tau \cdot u') \cdot (u_1 \circ x) = \rho(yx) \cdot u_1$. Thus, we have

$$F(\tau \cdot (x + u + u')) = \beta(u, u_1)\beta(u_1 , \rho(yx) \cdot u_1) - \beta(u, \rho(yx) \cdot u_1).$$

Now, we have $xy + yx = B(x, y) \cdot 1$, whence

$$\beta(u, \rho(y)\rho(x) \cdot u_1) = B(x, y)\beta(u, u_1) - \beta(u, \rho(x)\rho(y) \cdot u_1).$$

Making use of formulas (1), (3), Section 4.2, we have

$$\begin{aligned} B(x, y) &= B(y, x) = \beta(y \circ u_1 , x \circ u_1) = \beta(\rho(y) \cdot u_1 , \rho(x) \cdot u_1) \\ &= \beta(u_1 , \rho(yx) \cdot u_1) \end{aligned}$$

and therefore

$$F(\tau \cdot (x + u + u')) = \beta(u, \rho(x)\rho(y) \cdot u_1),$$

but $\rho(y) \cdot u_1 = y \circ u_1 = u'$, and

$$F(\tau \cdot (x + u + u')) = \beta(u, \rho(x) \cdot u') = F(x + u + u'),$$

which proves that τ leaves F invariant.

Set $\theta = \tau\mu(x_1)\tau^{-1} = \tau\mu(x_1)\tau$ and $\theta' = \mu(x_1)\tau(\mu(x_1))^{-1} = \mu(x_1)\tau\mu(x_1)$. These two operations are of order 2 and leave Ω and F invariant. We shall prove that they are identical. Let x be in M; then we have

$$\begin{aligned} \theta \cdot x &= \tau(\mu(x_1) \cdot (u_1 \circ x)) = \tau(\rho(x_1)\rho(x) \cdot u_1) \\ &= \beta(u_1 , \rho(x_1)\rho(x) \cdot u_1)u_1 - \rho(x_1)\rho(x) \cdot u_1 . \end{aligned}$$

We have $x_1x + xx_1 = B(x, x_1) \cdot 1$ and

$$\begin{aligned} B(x, x_1) &= B(x_1 , x) = \beta(x_1 \circ u_1 , x \circ u_1) \\ &= \beta(\rho(x_1) \cdot u_1 , \rho(x) \cdot u_1) = \beta(u_1 , \rho(x_1x) \cdot u_1), \end{aligned}$$

whence $\theta \cdot x = \rho(x)\rho(x_1) \cdot u_1$. On the other hand, we have

$$\begin{aligned} \theta' \cdot x &= \mu(x_1)\tau \cdot (\chi(x_1) \cdot x) = \mu(x_1) \cdot (u_1 \circ x_1xx_1^{-1}) \\ &= \mu(x_1) \cdot (\rho(x_1xx_1^{-1}) \cdot u_1) = \rho(x_1)\rho(x_1xx_1^{-1}) \cdot u_1 \\ &= \rho(xx_1) \cdot u_1 , \end{aligned}$$

since $x_1^2 = 1$; thus, θ' coincides with θ on M. On the other hand, we verify immediately that θ and θ' both map M into S_p : τ maps M into S_i, $\mu(x_1)$ maps S_i into S_p, and τ maps S_p into itself, whence $\theta(M) \subset S_p$; $\mu(x_1)$ maps M into itself, τ maps M into S_i, and $\mu(x_1)$ maps S_i into S_p, whence $\theta'(M) \subset S_p$. Since θ and θ' are of order 2 and coincide with each other on M, they coincide with each other on S_p, and $\theta\theta'$ coincides with the identity on $M + S_p$. On the other hand, θ and θ' belong to the orthogonal group of the quadratic form Ω, and so does $\theta\theta'$; since $\theta\theta'$ maps $M + S_p$ into itself, it maps also into itself the conjugate S_i of $M + S_p$ with respect to the associated bilinear form Λ of Ω. Since θ and θ' leave Ω and F invariant, so does $\theta\theta'$, and $\theta\theta'$ is an automorphism of the algebra A. Thus, it follows from IV.2.4 that $\theta\theta' = \mu(s)$ for some $s \in \Gamma_0$. Since $(\theta\theta')(x) = x$ for $x \in M$, s belongs to the kernel of the vector representation of Γ, whence $s = c \cdot 1$ for some $c \in K$. Since $\mu(s) \cdot u = cu = u$ for $u \in S_p$, we have $c = 1$ and $\theta\theta'$ is the identity, whence $\theta = \theta'$. Writing that $\theta'\theta$ is the identity mapping I, we obtain

$$\mu(x_1)\tau\mu(x_1)\tau\mu(x_1)\tau = I.$$

Let $J = \mu(x_1)\tau$: then $J^3 = I$. It is clear that J leaves F and Ω invariant, maps M onto S_p, S_p onto S_i, and S_i onto M; IV.3.1 is thereby proved.

It is clear that J is an automorphism of the algebra A. Making use of this automorphism, we obtain more formulas on the law of composition $\circ$. It follows from IV.2.3 that $Q(J \cdot x \circ J \cdot u) = \gamma(J \cdot x)\gamma(J \cdot u)$ if $x \in M$, $u \in S_p$, whence

$$Q(u \circ u') = \gamma(u)\gamma(u') \qquad (u \in S_p,\ u' \in S_i). \tag{2}$$

Since $x \circ (x \circ u) = Q(x) \cdot u$, we have

$$J \cdot x \circ (J \cdot x \circ J \cdot u) = \gamma(J \cdot x) J \cdot u,$$

$$J^{-1} \cdot x \circ (J^{-1} x \circ J^{-1} \cdot u') = \gamma(J^{-1} \cdot x) J^{-1} u',$$

whence

$$u \circ (u \circ u') = \gamma(u) u', \quad u' \circ (u \circ u') = \gamma(u') u \qquad (u \in S_p,\ u' \in S_i). \tag{3}$$

Applying J and J^{-1} again, we find the formulas

$$u \circ (u \circ x) = \gamma(u) x,\ u' \circ (u' \circ x) = \gamma(u') x \quad (x \in M,\ u \in S_p,\ u' \in S_i). \tag{4}$$

Making use of formula (3), Section 4.2, we obtain

$$B(u_1 \circ u', u_2 \circ u') = \gamma(u')\beta(u_1, u_2) \qquad (u_1, u_2 \in S_p,\ u' \in S_i),$$

$$B(u \circ u'_1, u \circ u'_2) = \gamma(u)\beta(u'_1, u'_2) \qquad (u \in S_p,\ u'_1, u' \in S_i). \tag{5}$$

Consider now the operation $\theta = \theta'$ introduced above. We have

$$J^{-1}\mu(x_1)J = \tau\mu(x_1)\mu(x_1)\mu(x_1)\tau = \tau\mu(x_1)\tau = \theta.$$

If $u \in S_p + S_i$, then we have $\mu(x_1)\cdot u = x_1 \circ u$, whence $\theta J^{-1}\cdot u = J^{-1}x_1 \circ J^{-1}u$. We have $J^{-1}x_1 = \tau\mu(x_1)\cdot x_1 = \tau \cdot x_1 = x_1 \circ u_1$; denote this element by u'_1. Then we see that $\theta\cdot v = u'_1 \circ v = v \circ u'_1$ if $v \in J^{-1}(S) = M + S_p$. On the other hand, we have $\mu(x_1)\cdot x = B(x_1, x)x_1 - x$ if $x \in M$. Thus, we obtain the formulas

$$\theta\cdot v = v \circ u'_1 \qquad (v \in M + S_p),$$

$$\theta\cdot u' = \beta(u', u'_1)u'_1 - u' \qquad (u' \in S_i). \tag{6}$$

On the other hand, τ is an automorphism of order 2 of A and $\tau\cdot x = x \circ u_1$ for $x \in M$. We have $(x \circ u_1) \circ u_1 = x$ by formula (4), whence $\tau\cdot(x \circ u_1) = (x \circ u_1) \circ u_1$. It follows that $\tau\cdot u' = u' \circ u_1$ for all $u' \in S_i$, and we have

$$\tau\cdot v = v \circ u_1 \qquad (v \in M + S_i),$$

$$\tau\cdot u = \beta(u, u_1)u_1 - u \qquad (u \in S_p). \tag{7}$$

4.4. Geometric Interpretation

Denote by Z and Z' two maximal totally singular subspaces of M, by u and u' representative spinors for Z and Z'.

Let x be a point of M. We know that a necessary and sufficient condition for x to be in Z is that $x \circ u = \rho(x)\cdot u = 0$ (III.1.4). Assume now that this condition is not satisfied. The space $Z + Kx$ is then of dimension 5; we shall see that this space contains exactly one maximal totally singular subspace $Z_1 \neq Z$ and that $\rho(x)\cdot u$ is a representative spinor for Z_1. The conjugate of Z is Z itself and therefore does not contain x, which shows that Z contains an element y such that $B(x, y) = 1$. Let $x_1 = x - Q(x)y$; then $Q(x_1) = 0$ and x_1 is not in Z. Let R be the space of elements $z \in Z$ such that $B(x, z) = 0$; R is of dimension 3, and $Z_1 = R + Kx_1$ is of dimension 4 and totally singular, because $Q(x_1) = 0$ and x_1 is orthogonal to every element of R, since both x and y are. The space R is in the conjugate of $Z + Kx$; if Z'_1 is any totally singular subspace of $Z + Kx$, then so is $R + Z'_1$, whence $R \subset Z'_1$ if Z'_1 is maximal totally singular. Moreover, Z'_1, which is of dimension 4, has an element $\neq 0$ in common with the 2-dimensional subspace $Kx + Ky$ of $Z + Kx$; if $Z'_1 \neq Z$, then y is not in Z'_1 and Z'_1 contains an element of the form $x + ay$, $a \in K$. Since $Q(x + ay) = 0$, we have $a = -Q(x)$ and $x_1 \in Z'_1$, whence $Z'_1 = Z_1$, which shows that Z_1 is the only totally singular subspace $\neq Z$ of $Z + Kx$. We have $\gamma(\rho(x)\cdot u) = Q(x)\gamma(u)$

$= 0$, which shows that $\rho(x)\cdot u$ is a pure spinor (IV.1.1). Since $y \in Z$, we have $\rho(y)\cdot u = 0$, whence $\rho(x)\cdot u = \rho(x_1)\cdot u$. We have $\rho(x_1)\cdot(\rho(x_1)\cdot u) = \rho(x_1^2)\cdot u = 0$, since $x_1^2 = 0$; it follows that x_1 belongs to the space of which $\rho(x)\cdot u$ is a representative spinor. If $z \in R$, then we have $B(z, x) = 0$, whence $zx + xz = 0$, and

$$\rho(z)\cdot(\rho(x)\cdot u) = -\rho(x)\cdot(\rho(z)\cdot u) = 0,$$

since $z \in Z$. It follows immediately that $\rho(x)\cdot u$ is a representative spinor for Z_1. Thus, we have proved the following statement:

IV.4.1. *Let u be a representative spinor for a maximal totally singular space Z and x an element of M not in Z. Then $x \circ u$ is a representative spinor for the unique maximal totally singular space $\neq Z$ contained in $Z + Kx$.*

Further, we observe that $u \circ (x \circ u) = \gamma(u)x = 0$ by formula (4), Section 4.3.

Assume now that $Z \cap Z'$ is of dimension 3. Then $Z + Z'$ is of dimension 5, and, if x is any element of Z' not in Z, then $u' = c\rho(x)\cdot u$, c a scalar, whence $u \circ u' = 0$. We shall now prove that the converse of this is true:

IV.4.2. *Let Z and Z' be maximal totally isotropic subspaces of M one of which is even and the other odd. Let u, u' be representative spinors for Z, Z'. Then a necessary and sufficient condition for $Z \cap Z'$ to be of dimension 3 is that $u \circ u' = 0$.*

We know already that the condition is necessary. Now, assume that $u \circ u' = 0$. We may assume that Z is even and Z' odd. Let the automorphism J have the properties of IV.3.1, and set $v = J(u)$, $y = J(u')$, whence $v \in S_i$, $y \in M$. We have $\gamma(v) = \gamma(u) = 0$, and v is pure; moreover, $v \circ y = J(u \circ u') = 0$, which shows that y belongs to the maximal totally singular space $\bar{Z}$ of which v is a representative spinor. Let x be an element of M such that $B(x, y) \neq 0$, Z_1 the maximal totally singular subspace of $\bar{Z} + Kx$ distinct from $\bar{Z}$, and v_1 a representative spinor for Z_1. Then, we have $v = ay \circ v_1$, a being a scalar $\neq 0$. We have $v_1 \in S_p$, whence $z = J^{-1}(v_1) \in M$ and $u = au' \circ z$. Making use of IV.4.1, we conclude that $\dim Z \cap Z' = 3$.

IV.4.3. *Let Z and Z' be maximal totally singular subspaces of M one of which is even and the other odd, and let u, u' be representative spinors for Z, Z'. If $u \circ u' \neq 0$, then* $\dim (Z \cap Z') = 1$ *and $u \circ u'$ is a basic vector of $Z \cap Z'$.*

We know that dim $(Z \cap Z') \equiv 1 \pmod 4$; since this dimension is $\neq 3$, it is 1. We have $u \circ (u \circ u') = \gamma(u)u' = 0$, $u' \circ (u \circ u') = \gamma(u')u = 0$; it follows that $u \circ u' \in Z \cap Z'$.

Let now $\bar{M}$ be the projective space whose points are the one-dimensional subspaces of M. Those one-dimensional spaces which contain singular vectors form a quadric hypersurface $\bar{Q}$ in $\bar{M}$. If $P \in \bar{Q}$, then any basic vector x of the subspace P of M will be called a representative vector for P. To the 4-dimensional totally singular subspaces of M correspond 3-dimensional projective subvarieties of $\bar{Q}$; if L corresponds to a 4-dimensional totally singular space Z, then any representative spinor for Z will also be called a representative spinor for L. The spaces L fall into two categories, corresponding to the two kinds of pure spinors; we shall denote by $\mathfrak{L}_e$ (respectively: $\mathfrak{L}_i$) the set of 3-dimensional projective varieties of $\bar{Q}$ whose representative spinors are even (respectively: odd). To the automorphism J of IV.3.1, there corresponds a mapping $\bar{J}$ which assigns to every point of $\bar{Q}$ a variety in $\mathfrak{L}_e$, to every variety in $\mathfrak{L}_e$ a variety in $\mathfrak{L}_i$ and to every variety in $\mathfrak{L}_i$ a point of $\bar{Q}$.

IV.4.4. *Let P and P_1 be distinct points of $\bar{Q}$. A necessary and sufficient condition for the line PP_1 joining P to P_1 to be on $\bar{Q}$ is that $\bar{J}(P)$, $\bar{J}(P_1)$ should meet each other. If L, L_1 are in $\mathfrak{L}_e$, a necessary and sufficient condition for L and L_1 to meet each other is that $\bar{J}(L)$, $\bar{J}(L_1)$ should meet each other.*

Let x and x_1 be representative points for P, P_1. A necessary and sufficient condition for the line PP_1 to be on $\bar{Q}$ is that $Kx + Kx_1$ be totally singular. Since x and x_1 are singular, this condition is equivalent to the condition that $B(x, x_1) = 0$. Now, we have $\beta(J \cdot x, J \cdot x_1) = B(x, x_1)$ and, if $u, u' \in S_p$, $\beta(J \cdot u, J \cdot u') = \beta(u, u')$. On the other hand, we know that, if Z, Z' are maximal totally singular subspaces of M and u, u' representative spinors for Z, Z', then a necessary and sufficient condition for $Z \cap Z'$ to be $\neq \{0\}$ is that $\beta(u, u') = 0$ (III.2.4); IV.4.4 follows immediately from these facts.

IV.4.5. *Let P be a point of $\bar{Q}$ and L a variety in $\mathfrak{L}_e$. A necessary and sufficient condition for P to belong to L is that $\bar{J}(P)$, $\bar{J}(L)$ should have a 2-dimensional projective variety in common.*

This follows immediately from IV.4.2.

4.5. The Octonions

Let us select once and for all an element $x_1 \in M$ such that $Q(x_1) = 1$ and an element $u_1 \in S_p$ such that $\gamma(u_1) = 1$. We shall set $u'_1 = x_1 \circ u_1$, whence $u'_1 \in S_i$, $\gamma(u'_1) = 1$.

Let x and y be in M. Then we have $x \circ u'_1 \in S_p$, $y \circ u_1 \in S_i$, and $(x \circ u'_1) \circ (y \circ u_1) \in M$. We set

$$x * y = (x \circ u'_1) \circ (y \circ u_1);$$

this formula defines a bilinear law of composition on $M \times M$, i.e., a structure of algebra on M. We shall call this algebra the *algebra of octonions.*

The element x_1 is the unit element for our law of composition, for $x_1 \circ u'_1 = x_1 \circ (x_1 \circ u_1) = u_1$ (by IV.2.3), $u_1 \circ (y \circ u_1) = y$ by formula (4), Section 4.4, and, similarly, $x_1 \circ u_1 = u'_1$, $(x \circ u'_1) \circ u'_1 = x$. Let x and y be in M. Then we have

$$Q(x * y) = Q(x)Q(y). \tag{1}$$

For, $Q(x * y) = \gamma(x \circ u'_1)\gamma(y \circ u_1)$ by formula (2), Section 4.3, and $\gamma(x \circ u'_1) = Q(x)$, $\gamma(y \circ u_1) = Q(y)$ by IV.2.3.

We have $x_1 \in \Gamma$; for any $x \in M$, set

$$\bar{x} = \chi(x_1)\cdot x = x_1 x x_1 = B(x, x_1)x_1 - x;$$

then $\bar{x}$ is called the *conjugate octonion* of x.

IV.5.1. *The mapping $x \to \bar{x}$ is an antiautomorphism of the algebra of octonions.*

In order to see this, we use the automorphisms τ, θ of A which were introduced in Section 4.3. Making use of formulas (6), (7), Section 4.3, we see that we may write

$$x * y = \theta\cdot x \circ \tau\cdot y.$$

The operation $\chi(x_1)$ extends to an automorphism $\mu(x_1)$ of A. We have $\theta = \mu(x_1)\tau\mu(x_1)$, $\tau = \mu(x_1)\theta\mu(x_1)$; since $\mu(x_1)$ is of order 2, we have $\mu(x_1)\cdot x * y = (\tau\cdot\bar{x}) \circ (\theta\cdot\bar{y}) = (\theta\cdot\bar{y}) \circ (\tau\cdot\bar{x}) = \bar{y} * \bar{x}$, which proves IV.5.1.

We shall now prove the formula

$$\bar{x} * (x * y) = Q(x)y \qquad (x, y \in M). \tag{2}$$

We have

$$\bar{x} * (x * y) = \theta\mu(x_1)\cdot x \circ (\tau\theta x \circ y)$$

because τ is an automorphism of order 2. Now, $\theta\mu(x_1) = \mu(x_1)\tau = \tau\theta$, since $\theta = \tau\mu(x_1)\tau$ (see Section 4.3). Thus

$$\bar{x} * (x * y) = \tau\theta(x \circ (x \circ \theta\tau\cdot y)) = \tau\theta\cdot(Q(x)\theta\tau\cdot y) = Q(x)y,$$

which proves (2).

If we replace $\bar{x}$ by its value $B(x, x_1)x_1 - x$ in (2), we obtain

$$B(x, x_1)x * y - x * (x * y) = Q(x)y.$$

On the other hand, it follows from (2) that $\bar{x} * x = Q(x)x_1$, whence

$$B(x, x_1)x - x * x = Q(x)x_1$$

and

$$B(x, x_1)x * y - (x * x) * y = Q(x)y,$$

which proves that

$$x * (x * y) = (x * x) * y.$$

Going over to the conjugates, we obtain, in virtue of IV.5.1, $(\bar{y} * \bar{x}) * \bar{x} = \bar{y} * (\bar{x} * \bar{x})$, or, since x, y are arbitrary, $(y * x) * x = y * (x * x)$. Thus, the difference

$$U(x, y, z) = x * (y * z) - (x * y) * z$$

is zero if y is equal to either x or z. Writing that $U(x, y + z, y + z) = 0$, we obtain $U(x, y, z) = -U(x, z, y)$, whence $U(x, y, x) = 0$. Thus, we see that $U(x, y, z) = 0$ whenever two of x, y, z are equal to each other. This is the characteristic property of what is called the *alternating algebras*. It implies that the algebra generated by any two octonions is associative, a fact which it is also easy to check directly.

We have defined the law of composition $*$ in M in terms of the law of composition of the algebra A. It is also possible to do the converse. Let J be the operation $\mu(x_1)\tau$; we know that J is an automorphism of order 3 of A, and every element of A is uniquely representable in the form $x + J\cdot y + J^2\cdot z$, where x, y, z are in M. Moveover, we have $J\cdot y = \mu(x_1)\tau y = \theta\cdot\bar{y}$, $J^2\cdot z = \tau\mu(x_1)z = \tau\cdot\bar{z}$, whence $Jy \circ J^2z = \bar{y} * \bar{z}$ and therefore $J^2y \circ z = J(\bar{y} * \bar{z})$, $y \circ Jz = J^2(\bar{y} * \bar{z})$, i.e.,

$$x \circ J\cdot y = J^2\cdot(\bar{x} * \bar{y}), \qquad x \circ J^2\cdot z = J(\bar{z} * \bar{x}), \qquad Jy \circ J^2z = \bar{y} * \bar{z}. \tag{3}$$

We shall now determine the automorphisms of the algebra of octonions. We have constructed in Section 4.2 a representation μ of the group Γ_0 by automorphisms of A. Let s be any element of Γ_0^+ such that $\mu(s)\cdot x_1 = x_1$, $\mu(s)\cdot u_1 = u_1$. Then we have also $\mu(s)\cdot u'_1 = \mu(s)\cdot x_1 \circ u_1 = u'_1$. If $x, y \in M$, then

$$\begin{aligned}\chi(s)\cdot x * y &= \mu(s)\cdot x * y\\ &= \mu(s)\cdot(x \circ u'_1) \circ (y \circ u_1)\\ &= (\chi(s)\cdot x \circ u'_1) \circ (\chi(s)\cdot y \circ u_1)\\ &= (\chi(s)\cdot x) * (\chi(s)\cdot y),\end{aligned}$$

which proves that $\chi(s)$ is an automorphism of the algebra of octonions. Let conversely σ be any automorphism of this algebra. Since x_1 is the neutral element, we have $\sigma \cdot x_1 = x_1$. We shall prove that $\sigma \cdot \bar{x} = \overline{\sigma \cdot x}$ for every $x \in M$. We have

$$Q(x)x_1 = \bar{x} * x = B(x, x_1)x - x * x,$$

whence

$$Q(x)x_1 = B(x, x_1)\sigma \cdot x - \sigma \cdot x * \sigma \cdot x,$$

but also

$$Q(\sigma \cdot x)x_1 = B(\sigma \cdot x, x_1)\sigma \cdot x - \sigma \cdot x * \sigma \cdot x,$$

and therefore

$$(Q(\sigma \cdot x) - Q(x))x_1 = (B(\sigma \cdot x, x_1) - B(x, x_1))\sigma \cdot x.$$

If x, x_1 are linearly independent, then so are $\sigma \cdot x$ and x_1, and $B(\sigma \cdot x, x_1) = B(x, x_1)$; this last formula is also obviously true if $x \in Kx_1$. Since $\bar{x} = B(x, x_1) - x$, it is clear that $\sigma \cdot \bar{x} = \overline{\sigma \cdot x}$. We may now extend σ to a linear automorphism $\bar{\sigma}$ of the vector space A by setting

$$\bar{\sigma} \cdot (x + J \cdot y + J^2 \cdot z) = \sigma \cdot x + J \cdot \sigma y + J^2 \cdot \sigma z.$$

Making use of the formulas (3), we see immediately that $\bar{\sigma}$ is an automorphism of A, and this automorphism maps M, S_p, and S_i onto themselves. It follows by IV. 2.4 that $\bar{\sigma} = \mu(s)$, where s is some element of $\Gamma^+{}_0$. Moreover, it is clear that $\bar{\sigma}$ commutes with J. We have $J \cdot x_1 = \mu(x_1)\tau \cdot x_1 = x_1 \circ (u_1 \circ x_1) = u_1$; since $\bar{\sigma} \cdot x_1 = x_1$, we have $\bar{\sigma} \cdot u_1 = u_1$, and $\mu(s)$ leaves u_1 fixed.

The automorphisms of the algebra of octonions may be characterized in still another manner. Let s be any element of Γ_0^+; then $\mu(s)$ is an automorphism of A, and so is $J\mu(s)J^{-1}$. It follows from IV.2.4 that $J\mu(s)J^{-1}$ may be written in the form $\mu(j \cdot s)$, where $j \cdot s$ is an obviously uniquely determined element of Γ_0^+; the mapping $s \to j \cdot s$ is an automorphism of order 3 of Γ_0^+. It is clear that $j \cdot (c \cdot 1) = c \cdot 1$ if c is an element $\neq 0$ of K; thus, j maps the kernel of the vector representation χ of Γ_0^+ into itself and defines an automorphism $\bar{j}$ of order 3 of the group $G_0^+ = \chi(\Gamma_0^+)$. We have seen above that an automorphism σ of the algebra of octonions may be written in the form $\chi(s)$, where s is an element of Γ_0^+ such that $\mu(s) = \sigma$ commutes with J; it follows that σ is an element of G_0^+ which is left invariant by $\bar{j}$. Conversely, let σ be any element of G_0^+ such that $\bar{j} \cdot \sigma = \sigma$. Write $\bar{\sigma} = \chi(s)$, where $s \in \Gamma_0^+$. We have $\chi(j \cdot s) = \chi(s)$, whence $j \cdot s = cs$, $c \in K$. Since j is of order 3, we have $c^3 = 1$; on the other hand, we have $\lambda(s) = 1$, $\lambda(j \cdot s) = 1$, whence $c^2 = 1$. It follows that $c = 1$ and that $\mu(s)$ commutes with J. Let x, y be in M, and $x' = \sigma \cdot x$, $y' = \sigma \cdot y$; then we have

$$\bar{x}' * \bar{y}' = J \cdot x' \circ J^2 y' = J\mu(s) \cdot x \circ J^2\mu(s) \cdot y$$
$$= \mu(s)J \cdot x \circ \mu(s)J^2 \cdot y = \sigma \cdot (Jx \circ J^2 y) = \sigma(\bar{x} * \bar{y})$$

and σ is an automorphism for the law of composition $(x, y) \to \bar{x} * \bar{y}$. This law of composition admits x_1 as its neutral element, whence $\sigma \cdot x_1 = \mu(s) \cdot x_1 = x_1$. Since $\mu(s)J = J\mu(s)$ and $J \cdot x_1 = u_1$, $\mu(s)$ leaves u_1 fixed, and $\sigma = \chi(s)$ is an automorphism of the algebra of octonions. Thus, *the group of automorphisms of the algebra of octonions is the group of elements of G_0^+ which are left fixed by the automorphism $\bar{j}$ of order 3 of this group.*

If x is an octonion, we set

$$\operatorname{Sp} x = B(x, x_1)$$

and call this element the *trace* of x. It is clear that

$$\operatorname{Sp} \bar{x} = \operatorname{Sp} x.$$

We shall now prove that

$$\operatorname{Sp} x * y = B(\bar{x}, y)$$

if x, y are octonions. The left side is

$$B(x * y, x_1) = \Lambda((x \circ u'_1) \circ (y \circ u_1), x_1) = \Phi(x \circ u'_1, y \circ u_1, x_1);$$

by the symmetry of Φ, this is also

$$\Phi(x_1, x \circ u'_1, y \circ u_1) = \Lambda(x_1 \circ (x \circ u'_1), y \circ u_1).$$

Now, we have

$$x_1 \circ (x \circ u'_1) = \rho(x_1)\rho(x) \cdot u'_1 = \rho(\bar{x})\rho(x_1) \cdot u'_1 = \rho(\bar{x}) \cdot u_1,$$

since $\rho(x_1) \cdot u'_1 = u_1$. Thus, $\operatorname{Sp} x * y = \Lambda(\bar{x} \circ u_1, y \circ u_1) = B(\bar{x}, y)$ by formula (3), Section 4.2.

Since $\chi(x_1)$ belongs to the orthogonal group of Q, we have $B(\bar{x}, \bar{y}) = B(x, y)$; it follows immediately that

$$\operatorname{Sp} y * x = \operatorname{Sp} x * y.$$

On the other hand, since B is nondegenerate, the same is true of the bilinear form $(x, y) \to \operatorname{Sp} x * y$.

Let x, y, z be octonions. Then we have

$$\mathrm{Sp}\,(x * y) * z = \mathrm{Sp}\, x * (y * z).$$

The left side is equal to

$$\begin{aligned} B(\bar{z},\, x * y) &= \Lambda(\mu(x_1)z,\ \theta\cdot x \circ \tau\cdot y) \\ &= \Phi(\mu(x_1)z,\ \theta\cdot x,\ \tau\cdot y) \\ &= \Phi(\tau\cdot z,\ \mu(x_1)\cdot x,\ \theta\cdot y) \end{aligned}$$

because the automorphism $\tau\mu(x_1)$ of A leaves Φ invariant and $\tau\mu(x_1)\theta = \mu(x_1)$, $\tau\mu(x_1)\tau = \theta$. Now we have

$$\Phi(\tau\cdot z,\ \mu(x_1)\cdot x,\ \theta\cdot y) = \Phi(\mu(x_1)\cdot x,\ \theta\cdot y,\ \tau\cdot z) = \mathrm{Sp}\, x * (y * z).$$

Since $\mathrm{Sp}\, y * x = \mathrm{Sp}\, x * y$, we see that

$$\mathrm{Sp}\,(x * y) * z = \mathrm{Sp}\,(y * z) * x = \mathrm{Sp}\,(z * x) * y.$$

We have

$$\mathrm{Sp}\,(z * x) * (y * z) = \mathrm{Sp}\,(x * y) * (z * z),$$

for, from the preceding formulas, the left side is equal to

$$\mathrm{Sp}\,(y * z) * (z * x) = \mathrm{Sp}\, y * (z * (z * x)).$$

But $z * (z * x) = (z * z) * x$, and the left side of our formula is equal to

$$\begin{aligned} \mathrm{Sp}\, y * ((z * z) * x) &= \mathrm{Sp}\,((z * z) * x) * y \\ &= \mathrm{Sp}\,(z * z) * (x * y) \\ &= \mathrm{Sp}\,(x * y) * (z * z). \end{aligned}$$

Replacing z by $z\ t$, $z + t$ in the formula we have just proved, we obtain

$$\mathrm{Sp}\,(t * x) * (y * z) + \mathrm{Sp}\,(y * t) * (z * x) = \mathrm{Sp}\,(x * y) * (t * z + z * t),$$

where x, y, z, t are any octonions.

ACKNOWLEDGMENTS

THE AUTHOR acknowledges with gratitude the generosity shown by the following in making the publication of the Bicentennial Editions and Studies possible: the Trustees of Columbia University, the Trustees of Columbia University Press, Mrs. W. Murray Crane, Mr. James Grossman, Mr. Herman Wouk, and friends of the late Robert Pitney who wish to remain anonymous.

INDEX

Bei Fragen zur Produktsicherheit wenden Sie sich bitte an:
If you have any questions regarding product safety, please contact:

Walter de Gruyter GmbH
Genthiner Straße 13
10785 Berlin
productsafety@degruyterbrill.com